The Nature of Matte

The Nature of Matter

Wolfson College Lectures 1980

EDITED BY
J. H. MULVEY

CLARENDON PRESS OXFORD

1981

Oxford University Press, Walton Street, Oxford OX2 6DP

OXFORD LONDON GLASGOW
NEW YORK TORONTO MELBOURNE WELLINGTON
KUALA LUMPUR SINGAPORE JAKARTA HONG KONG TOKYO
DELHI BOMBAY CALCUTTA MADRAS KARACHI
NAIROBI DAR ES SALAAM CAPE TOWN

Published in the United States by Oxford University Press, New York

British Library Cataloguing in Publication Data

The Nature of matter.
1. Matter—Addresses, essays, lectures
I. Mulvey, J. H.
530 QC171.2

ISBN 0-19-851151-5

Printed and bound in Great Britain
at The Pitman Press, Bath

Contents

Preface

The Nature of Matter is based on the eight Wolfson Lectures on this theme given at Wolfson College, Oxford, in the Spring of 1980. The aim of the series was to tell a general, non-specialist audience about the remarkable progress made over the past ten to fifteen years in the search for an understanding of the basic components of matter and the forces determining their behaviour; not since the early decades of the century have there been so many revolutionary discoveries and such a rush of new ideas. All the speakers have made major contributions to our understanding of matter and the College Hall overflowed for every talk.

It was a great pleasure to organize the series and I would like to thank the speakers, again, for their lectures and for the considerable effort of writing them up. Any errors that have occurred in the course of editing the lectures into a book are my responsibility.

I would like to thank the President and Fellows of Wolfson College for their support and encouragement and I am especially grateful to Gillian Moore who looked after all the administration so well. I should also thank Paul Boddington and all the College staff who helped with the arrangements for each lecture, my colleagues in the Department of Nuclear Physics who arranged the closed-circuit TV for the overflow rooms, Tricia Falconer Smith for typing several of the manuscripts, and Alan Holmes for the cartoons in chapters 1 and 5.

Permission to use the following is gratefully acknowledged:
Fig. 1.2 Apollo photograph provided by the National Space Science Data Centre through the World Data Centre A for Rockets and Satellites. Fig. 3.2 A drawing by M. C. Escher, © SPADEM Paris, 1981. Fig. 3.5 *Superman* illustration copyright © 1938 Detective Comics Inc. renewed © 1966, D. C. Comics Inc. Fig. 8.5 Einstein cartoon, © Sidney Harris, American Scientist Magazine. Other sources are acknowledged on the illustrations.

Wolfson College, Oxford
November 1980

J.H.M.

'If therefore we consider the . . . natural philosophers, . . . with regard to their knowledge of phenomena, we shall find it consists . . . in a greater largeness of comprehension, whereby analogies, harmonies, and agreements are discovered in the works of nature, and the particular effects explained, that is, reduced to general rules; which rules . . . are most aggreeable and sought after by the mind; for that they extend our prospect beyond what is present and near to us, and enable us to make very probable conjectures touching things that may have happened at very great distances of time and place, as well as to predict things to come; which sort of endeavour towards Ommnisence is much affected by the mind.'

Bishop Berkeley, *The Principles of Human Knowledge* (1710).

Introduction

JOHN MULVEY

Looking in a mirror, seeing a diamond sparkle, wondering about the Sun and the stars often prompts one to question the nature of matter. The belief that the infinite variety of the physical world can be reduced to a finite, even small number of things and ideas was a challenge accepted by ancient philosophy and is today the motivation for that branch of fundamental research usually referred to as 'high-energy' or 'particle' physics. These names of convenience seem sometimes to give the misleading impression that this research is pursued somehow off the main stream of scientific endeavour as a race between rival communities of physicists to reach the highest energies by building the largest (and so most expensive) particle accelerators, or as an esoteric form of 'stamp collecting' with rewards for finding ever more rare types of particle all of dubious relevance in the 'real world'. So the first point I wish to discuss is the context in which we should see particle physics and thus the main theme of the lectures collected in this book.

Research in physics covers a very wide range of phenomena. It underlies and penetrates most other sciences including chemistry, biology, genetics, astronomy, astrophysics, metallurgy, and engineering, providing basic concepts and new techniques. Much of the present research in physics is devoted to the study of assemblies of atoms and molecules, with special emphasis on details of internal structure or the surfaces of solids, and often in conditions of precisely controlled composition, high pressure or extremely low temperature, and so on. The rapidly increasing understanding of matter, in macroscopic forms, obtained in this research over the years has fed the continuous advance of modern technology; for example, in electronics it has led to revolutionary developments in communications, computing, the micro-chip, and robotics, all carrying dramatic implications for the way we live. This research into the macroscopic states of matter has its foundations in the discoveries made in the first quarter of this century when the atom and quantum mechanics formed the frontier of basic physics. Today that frontier is at the particle level, below atomic or nuclear scales, and experiments using particles of high energy are necessary to pursue the search for an understanding of the nature of matter. We cannot predict what use, if any, might one day be made of this deeper understanding of Nature or if it will have any impact at all on the future evolution of society. But our characteristically restless curiosity will continue to ask how the world came to be as it is, what is the nature of the forces which shape its destiny, what the electrons, protons, and neutrons making up the atoms are, and what may be their fate? These are among the most fundamental questions that may be approached through observation and experiment guided by the genius of human imagination and have a deep appeal to the intellect; we are children of the Universe, its historians, and its prophets.

It is unfortunate, but inevitable, that the language developed to describe everyday happenings is not adequate when we probe deeper into these mysteries. This is a problem common to most scientific research but in physics the communication of a full understanding of what has been learned requires an apprenticeship of several years: the language is essentially mathematical and one must also grasp a number of abstract concepts many of which have no parallel in the macroscopic world of common experience. It is a wonder that a coherent description can be found for such a vast range of phenomena; as Einstein remarked: 'The most incomprehensible thing about the Universe is that it is comprehensible'. But it is not necessary to be a physicist to appreciate the significance of the discoveries made by those who inquire into the nature of matter. Many enjoy Beethoven's quartets without being able to read the score or play a note. The achievements of science are, like great music, literature, drama, and art, triumphs of the human mind—sometimes slow to be appreciated and requiring some effort of comprehension but eventually becoming part of the cultural heritage. In the Spring of 1980 eight 'natural philosophers' came to Wolfson College to talk to lay audiences about the present view of the nature of matter and to explain the significance of some of the dramatic discoveries of recent years. Their aim was to convey the essence of these discoveries, how they have led to new perceptions of matter and the forces controlling the Universe; to report on the current state of research at this frontier of science, to outline the crucial questions for future experiments and how we might progress towards a more complete understanding of all physical phenomena. The lecturers interpret for us 'the shadows cast on the walls of the cave' by events taking place in the laboratories of high-energy physics or in distant regions and epochs of the Universe.

The first two chapters are introductory but all the chapters stand largely on their own. This leads naturally to some repetition, but each author approaches the main theme from a different angle, and with different emphasis brings his own insight to illuminate the central questions.

The series is opened by Sir Denys Wilkinson with a *tour d'horizon*, establishing the range and scale of the particles and forces and the Universe in which they hold sway, and touching on many of the points to be explored in more detail in the succeeding chapters. Sir Rudolf Peierls takes us further in the discussion of particles and forces, introducing some of the odd laws which govern behaviour in the sub-atomic world and which determine the way forces work to give matter its various characteristics. He outlines the recent history of the search for elementary constituents of matter in this century and the replacement of the simplicity anticipated in the 1930s by the proliferation of particle states uncovered in the 1950s and 1960s which led to the quark model.

Chris Llewellyn Smith introduces the concept of symmetry as applied to the laws of Nature in chapter 3. There is, among physicists, a deep-seated faith that the laws of Nature are not only not arbitrary but that they should be expressible as a concise and structurally elegant set of principles. The full power of symmetry principles is realized in quantum field theory where they are found to determine the structure of the theory even when their presence is hidden from direct view by, it seems, a conspiracy of influences biasing the choice made by Nature towards an asymmetric solution for the world we observe. This notion, that the phenomena accessible to observation need not directly reflect underlying symmetry principles, is a novel departure in thinking which has opened the way to fresh and ambitious attempts to formulate theories unifying our understanding of the forces of Nature in spite of their apparent diversity.

In chapter 4 Donald Perkins reviews the experiments of the last decade or so which have established the quarks as constituents of the protons and neutrons, a level of sub-structure which had already been strongly suggested by the very specific patterns formed by the many particle states found in earlier years. He also outlines the present view of the force responsible for binding the quarks together and which appears to have the unique property of never relinquishing its grip, so that quarks remain permanently confined within the particles they constitute. This is another novel situation; for the first time we accept the existence of entities which we may never be able to verify by direct observation in the laboratory.

Abdus Salam recounts, in chapter 5, the history of his progress towards a theory bringing the electromagnetic and weak (radioactivity) forces together in one framework. For his contributions to this theory he received a share, with Glashow and Weinberg, of the 1979 Nobel Prize for Physics. This is a tale full of insights into the development of a new theory, as it progresses partly by analogy, by seeing new connections between apparently dissimilar phenomena, by building on foundations laid earlier and sometimes with new ideas squashed by 'elder statesmen' of the physics establishment. The crux of the approach is the idea brought out in chapter 3 that the fundamental laws of physics may respect symmetry principles not manifest in the present conditions of the Universe. Abdus Salam goes on to outline some of the ideas underlying current attempts to find a single unifying principle for all the forces of Nature, including gravity, and embracing the elementary forms of matter. This is the frontier of present thought where as yet speculation and conjecture go largely unconstrained by experiment. Indeed the conditions of matter required to test some predictions of these theories are unattainable in the laboratory, and, since the 'Big Bang', may never again occur in the Universe.

This leads us to chapter 6 and a discussion by John Ellis of the 'cosmological connection', a number of points of contact between particle physics and cosmology where one discipline illuminates the other. Nuclear and particle physics provide the basis for an understanding of many astrophysical processes and the extreme situations which are relatively commonplace in a violent and almost infinite Universe offer a means of studying the behaviour of matter in un-Earthly conditions. No place or time is more extreme than the Big Bang, now generally accepted as the origin of our Universe, and, with signs of a success as incredible as the audacity of the attempt, the still speculative theories which attempt to unify the forces are drawn upon to construct, for the first time, a plausible explanation of the existence of matter: as a relic of the Big Bang, the unpaired 'wallflowers', as Ellis puts it, at the first (and nearly the only) 'dance'.

The tools of research described by John Adams in chapter 7 are engineering marvels put together by close collaborations of physicists and engineers (and physicists turned engineers) successfully expanding the boundaries of technology to achieve higher performance and to develop new methods. The discoveries of high-energy physics today would be quite impossible without the fullest exploitation of the technology based on the fundamental research of yesterday. The machinery of experimental particle physics is huge, expensive and, in Europe at least, largely dependent on international co-operation in the support of CERN, the centre for research near Geneva. The experimentalists who choose this field have to join forces in large international collaborations with other physicists and engineers to mount their complex experiments. They must design their apparatus not only to meet the demands of their experiment but also to minimise cost in the production of the large number of components, to meet firm delivery dates set by the schedule of the accelerators and to run reliably over many months or years. These university staff and research students have to learn to work in an environment more akin to a highly professional modern industry than to the tranquility of academic life.

The last chapter is a summing-up by Murray Gell-Mann, who was awarded the Nobel Prize in Physics in 1969 for his many outstanding contributions to particle physics, including the idea of quarks. He distinguishes what he believes is established from what is as yet unsupported conjecture awaiting the verdict of experiment, and lists a number of the crucial questions now confronting us.

A glossary has been included to help with unfamiliar words or ideas, to summarize some of the definitions, and to enlarge on some of the explanations given in the text. Lastly, comes a Bibliography. No references are given in the main text but some may be found in the specialist review articles listed here.

Finally, I hope this book will convey some of the present atmosphere of excitement existing among those searching for an understanding of the nature of matter. The latest experiments and ideas seem to carry a promise, as seldom before, of a grand synthesis of all physical phenomena in a few succinct principles. But it is still only a promise and Nature has always been a match for the hubris of the scientist. The one lesson of history in this search is that there are surely surprises in store for those who expect the next rounds of experiments safely to confirm their preconceptions. And if the picture unfolded in these chapters seems too bizarre to be credible, remember the remark attributed to the great physicist, Niels Bohr, who told Pauli, after the latter had lectured on a new idea: 'We are all agreed that your theory is crazy. The question which divides us is whether it is crazy enough.'

1

The Organization of the Universe

SIR DENYS WILKINSON

Curiosity about the influences that shape the Universe of which we are part, about its fine structure and the ultimate constituent particles of matter, if such exist, is not by any means new in the thoughts of Man. The so-called atomists of the fifth century B.C., Leucippus and Democritus, had ideas that, cast into modern idiom, we could recognize as the germs of the way in which we now think about matter. But it is with Epicurus, as relayed to us from 'that peaceful garden' by Lucretius, that we first find a grasp of the idea that the apparent complexity and infinite variety of the visible world may hide an inner simplicity of structure existing at levels beneath our perception. Epicurus lived between the fourth and third centuries; Lucretius must have lived until at least 55 B.C., because he wrote disparagingly of the English climate in his great Epicurean work *De Rerum Natura*, which we might otherwise consider as an objective reflection upon Nature, morals, and the senses. I shall look to Lucretius for occasional comments upon my story and cannot do better than to begin by bidding you, through him:*

> Give your mind now to the true reasoning I have to unfold. A new fact is battling strenuously for access to your ears. A new aspect of the Universe is striving to reveal itself. But no fact is so simple that it is not harder to believe than to doubt at the first presentation.

Forces

The influences that shape our Universe are: the particles out of which our world is made, the forces between those particles, and the laws governing the play of those forces. We now recognize, phenomenologically, in our present-day Universe, four apparently distinct forces:

* The quotations from *De Rerum Natura* follow R. E. Latham's translation published by Penguin Books.

(a) *gravity*, which acts indiscriminately and attractively between all particles;

(b) the electric force that acts by definition only between electrically charged particles and that can be attractive or repulsive depending on whether the particles in question have opposite or similar charges; with electricity goes magnetism, generated by moving electric charges, as we have understood since the days of Faraday and Maxwell over a century ago, and the two are collectively called *electromagnetism*;

(c) the *strong*, or nuclear, force that acts only within a class of particles called hadrons, of which the neutrons and protons out of which the nucleus of the atom is made are examples, and which is indiscriminate in respect of the electrical charges of those particles;

(d) the *weak* force, which acts within another class of particles called leptons, of which the electron is an example; its action is independent of electric charge but it also acts between leptons and hadrons, and between hadrons and hadrons.

These four forces of Nature differ also in some of the laws that they obey: all the forces appear to conserve energy and momentum for example, but whereas the electromagnetic and strong forces show no intrinsic preference as between left-handed and right-handed phenomena (phenomena of a corkscrew nature such as circularly-polarized light or the spinning of a particle with respect to its direction of motion) the weak force does not respect this plausible symmetry; it shows a strong prejudice and can carry out certain operations only right-handedly and others only left-handedly. Thus in the process of radioactive beta-decay the electron that is emitted spontaneously from a neutron under the influence of the weak force, converting the neutron into a proton, comes out preferentially as a left-handed corkscrew; that is to say, as it departs it is spinning anti-clockwise as viewed by the proton that it leaves behind. The concept of symmetry and the behaviour of the forces under different kinds of symmetry transformation, such as replacing left-handedly spinning particles with right-handed ones, are of fundamental significance, as Chris Llewellyn Smith explains in chapter 3.

The four forces also differ in the way in which their effect decreases as the distance between the particles between which they act increases: the gravitational and electrical forces drop off only rather slowly with separation—the famous inverse square law of force—whereas the strong force falls off much more sharply than this and has a reach of only about 10^{-13} cm, beyond which it very rapidly becomes negligible; the weak force is also of short range in this same sense—we believe that its reach is very much less even than that of the strong force. These differences in range and strength may be illustrated by considering the force between two hydrogen atoms as a function of their distance apart. Each atom is a proton with an electron going around it at a distance of about 1 Å (or 10^{-8} cm) and we will take as a measure of separation the distance between the protons at their centres. At a separation of 1 fm (or 10^{-13} cm), a distance characteristic of the finite reach of the strong force, the strong force is about 100 times stronger than the electric force. The weak and gravitational forces are negligible in strength at this separation (although the effect of the weak force can be picked out, feeble as it is, by its tell-tale corkscrew asymmetry even within systems like atomic nuclei that are dominated by the strong force). As the strong force falls off much more rapidly with distance than the electrical force, at a separation of only about 5.6 fm the two are equal (the electrical force in question at this stage is that between the two protons, the circling electrons having negligible effect). At a separation of 1 Å, characteristic of atomic sizes and of atomic separations in molecules, the electrical force is totally dominant, but as the separation increases beyond that the electrical force itself drops off more rapidly than with the inverse square of the distance between the centres of the atoms. This is because, as the two hydrogen atoms separate, the negatively charged electrons come between the positively charged protons and, progressively more rapidly, cancel out or screen the electrical proton–proton force. But the gravitational force, minute as it is, depends only on the masses of the interacting particles and so eventually comes into its own, equalling the electrical force at about 2 mm separation and completely dominating at a few centimetres. Thus although the gravitational force is exceedingly weak (a hydrogen atom held together by gravity alone would be about as big as the entire visible Universe) it utterly dominates in large-scale circumstances of electrical neutrality and holds together the chief structures of the universe—the planets, the stars, the galaxies, the galactic clusters. . . . We are ourselves held onto our Earth by the gravitational force because both we and the Earth are electrically neutral, but if there were suddenly to be brought about an electrical imbalance of only 1 part in 10^6 between protons and electrons, the resultant repulsive force between each one of us and the Earth would be equivalent to about 10^{36} tons weight, which is a measure of the strength of the electrical force relative to the gravitational.

So there appear to be tremendous differences between these four forces of Nature, yet the possibility of there being a fundamental unity between them will be a major theme running through this book and, if there is, this may have dominated events in the first instants of the Universe. But before following up that idea, let us consider a point of philosophical importance: why these four forces? If we take an anthropocentric viewpoint we can see that we 'need' the gravitational force to hold together the stars, to hold the planets in their orbits around the Sun and to hold us onto the Earth; we 'need' the electrical force because it is that which holds the electrons onto their atoms and that holds the atoms together into the molecules out of which we are made; we 'need' the strong force because it is that which holds neutrons and protons together to give the variety of atomic nuclei and hence the range of atomic, and so chemical species essential for building the rich diversity of molecules on which life depends (life could scarcely have evolved from hydrogen alone). But why do we need the weak force? Very simply because it is that which initiates the fuelling of the stars and permits the building of the elements. Stars shine, generate the radiance upon which life depends, because when two protons in the hydrogen of a star bump up against one another, very occasionally one of those protons, in the course of collision, transforms itself through the weak force into a neutron, emitting a positive electron (the antiparticle of the familiar negative electron) and also some energy; the resultant neutron and the other proton stick together as a deuteron, the nucleus of 'heavy hydrogen'. Following this, other, much more rapid nuclear reactions depending not upon the weak but upon the strong and electromagnetic forces lead to the formation of alpha-particles, the nuclei of helium atoms containing two protons and two neutrons, with a considerable release of energy that provides most of the Sun's power. Similarly it is the weak force, with its ability to interconvert neutron and proton through radioactive beta-decay, that permits the building up of heavier atomic nuclei in stars by the addition of neutrons to lighter nuclei, followed, on occasion, by the conversion of neutron to proton inside the nucleus; this increases the positive electrical charge of the nucleus and so it is moved higher up the atomic table of the chemical elements. All four forces are essential if we are to be here to know about them.

There is also another, subtler, condition for our existence which is to do with the balance, within the Universe, of matter and radiation—light of all wavelengths: this balance is about 10^9 to 1 in favour of radiation as measured in terms of numbers of photons, the units of radiant energy, to numbers of neutrons and protons. For reasons to do with the degree of order in the Universe, if this balance were very different we should not be here. Just why these delicate provisions of forces and conditions, critical to our existence, have been made I leave for others to debate.

The tremendous differences between the four forces of Nature have encouraged, rather than discouraged, the quest to find out how, if at all, those forces are interrelated. Are they indeed totally unrelated, each *sui generis*, or can they be linked through some principle of unification so that, in some sense, they are different aspects of each other? Some have believed that they must be related, others that they cannot be related. Einstein spent the last decades of his life unsuccessfully attempting to unify electromagnetism and gravitation; the former beautifully and probably completely (at least in the classical, pre-quantum-mechanical sense) described by Maxwell's equations; the latter beautifully and possibly also completely (classically) described by Einstein's own equations of general relativity. These unsuccessful attempts were generally regarded as unprofitable exercises; as Wolfgang Pauli remarked: 'What God hath put asunder let no man join'. Nevertheless this attempt to unify electromagnetism and gravity was perhaps natural in view of the inverse square law for the dependence on distance of both the gravitational and electrical forces; but we now understand why the difficulties of that possible eventual unification are greater than those facing the possible unification of the other forces.

Indeed, it now appears, thanks to the work chiefly of Abdus Salam and Steven Weinberg, that the electromagnetic and weak forces have been unified: in 'everyday life' there appear to be two, totally different, forces but they are really just two aspects of a single, more fundamental, interaction. Above a certain régime of energy of interaction between particles, at present unattainable in the laboratory but not totally out of practical reach, the two forces will become of the same strength and will merge into a single force, the electro-weak force, acting equally and indiscriminately between all particles, hadrons and leptons, electrically-charged and electrically-neutral. Hadrons and leptons will still remain distinguished, as before, by the strong force which, at the energy of interaction at which the electric and weak forces merge, still acts only between hadrons and which, with gravity, remains in the wings. The scale of the electro-weak unification energy is about 100 GeV, about 100 times the mass of the proton, so that at energies somewhat above this the two forces fuse fully into the one.

It seems as though the strong force may now, in fact, be emerging from the wings to join the other two. We are in a stage of rich and attractive speculation about what are generically known as Grand Unified Theories (GUTs) that combine the electro-weak and strong forces such that in some yet higher energy régime than 100 GeV the distinction between strong and electro-weak forces is lost, just as above 100 GeV the weak and electromagnetic forces come together. It seems that this GUTs energy régime must be very high indeed: of the order of 10^{15} GeV, 10^{15} times the mass of a proton.

Of the possible ultimate unification of gravity with the other force, in the sense of GUTs, or with the other three forces in the 'everyday' sense, we can only speculate rather blindly although such speculations, the so-called 'super-symmetric' theories, already abound. A chief reason for our difficulty over gravity is that we do not, as yet, have an acceptable quantum theory for it; indeed it is not yet demonstrated that gravity is basically of a quantum nature at all and if it is not it will not be unifiable with the other force or forces without an understanding of a type that we do not yet possess; for the time being the veil is drawn. This is by no means to say that gravity cannot have quantum effects and in fact such are the basis of the mechanism by which Stephen Hawking has described the 'evaporation of black holes'. It is also clear that if we attempt to discuss concentrated masses of greater than a certain magnitude called the Planck mass of about 10^{19} GeV we shall simply not know how to do it because for such masses the quantum *effects* of gravity will dominate the situation, no matter how gravity itself is structured in the quantum sense. For such masses the gravitational force between particles is comparable to the forces of GUTs, the strong and electro-weak forces; thus before such conditions of matter can be considered the problem of the relationship between all the forces, or super-unification, must be understood.

Alpha and omega

We are now ready to take a journey in time, back towards the earliest instants of time, if such in any comprehensible sense there were following the 'Big Bang' of instantaneous creation. The basis of most current discussion of the Universe is the Big Bang model which says that in the beginning there was a great concentration of energy, finite in amount but infinite in density, from which the Universe evolved according to immutable Laws of Nature. If this model is wrong and there was not a Big Bang in this sense then ignore everything which follows about the first few minutes of the Universe, but the rest stands. It is also assumed that time and space, the coordinates of the Laws of Nature, are infinitely sub-divisible and do not, themselves, have some sort of ultimate granularity. So far there are no signs of such granularity appearing and, on the contrary, to the finest scales of space and time yet probed (about 10^{-16} cm and 10^{-26} seconds) all seems completely smooth and well. But let it also immediately be said that the times and distances we shall discuss with apparent confidence are vastly beyond this present experimental range of verification and so it could all be vitiated by a granularity not yet perceived. However, our discussion of times longer than 10^{-26} seconds and distances greater than 10^{-16} cm is probably reliable.

At an age of about 10^{-45} seconds the typical energy of constituents of the embryo Universe would have been in the region of 10^{19} GeV, the Planck mass. Earlier than this our primitive understanding, in the absence of the final unification of the four forces, dissolves into an indescribable seething foam of mini black holes, somehow forming, evaporating, and reforming; we are in the realm of total speculation, no longer confident of the distinction of time and space and we must, at least until the great steps to final unification are taken, be silent. Even after the Planck time has elapsed the Universe is not all plain sailing from the point of view of our present physical understanding. In particular, we must understand why our Universe today consists almost solely of particles, with very few antiparticles. In the beginning, unless the Creator wished to impose special conditions upon His creation, there was simply energy: a finite amount in all probability, but infinitely concentrated. Now when, in today's laboratories, we simply let energy loose through the violent collision of accelerated protons, for example, new particles and their antiparticles are created together out of that energy, following Einstein's $E = mc^2$, in equal abundance. Why, then, is the Universe itself, also, as we rather confidently speculate, made originally just out of energy, so lop-sided in the particle versus antiparticle sense? We must assume that at the end of the Planck time, after the eventual lapse of those 10^{-45} seconds, the evaporation of all, or most of, the black holes gave us, in a way that we do not yet know how to discuss, the beginnings of our modern Universe with equal numbers of particles and antiparticles. However these were not the particles and antiparticles found in today's Universe, neutrons, protons, electrons, etc., but rather progenitors of those particles: progenitor particles whose own radioactive decay eventually gave rise to the building materials of the edifices—including ourselves—of our present interest. These are new ideas emerging from the attempts to construct GUTs. The progenitor particles entered the scene after gravity slackened its grasp with masses at the GUTs unification energy of about 10^{15} GeV. In chapter 6 John Ellis explains how, starting with equal numbers of progenitor particles and antiparticles, the present preponderance of one sort of matter over the other came into being; a catastrophic self cancelling annihilation of equal amounts of matter and antimatter, creating a universal flood of radiation, was averted, leaving the observed, 'needed', ratio of about 10^9:1 between radiation and matter by the time the Universe was 10^{-35} seconds old.

All this does not, of course, prove that GUTs are correct, but it does suggest that we are not necessarily begging any fundamental questions in dating the modern history of the evolution of our Universe from about 10^{-35} seconds after Creation, complete with the implied preponderance of particles over antiparticles and with the implied balance of radiation as against matter.

After the first 10^{-35} seconds we believe things took a relatively conventional course based on physics in which we can feel greater confidence, better grounded in observational fact, many unknowns and uncertainties as there still may be. One major uncertainty concerns the very internal structures of the neutron and proton themselves and of all other hadrons; a related uncertainty is the nature of the strong force. So far we have spoken of the strong force as though, in its essence, it acts *between* hadrons; however, the basic strong force is more likely to be that acting *within* hadrons, that *between* hadrons being no more than an external reflection of this. The direct descendants of the progenitor particles are thought to be not the neutrons, protons, and the other hadrons themselves but rather the *constituents* of those particles, the quarks, between which the true basic strong force has its play; but more of this later.

The GUTs may offer an explanation for the origin of the matter we are made of but it seems difficult to take seriously a theory of events occurring at a time of zero plus 10^{-35} seconds we can scarcely imagine and at an energy of 10^{15} GeV that we can never achieve in our laboratories. (An accelerator wrapped around the entire equator of the earth and equipped with the most powerful magnetic field we know how to make to guide its particles would generate an energy of only about 10^7 GeV; such an accelerator for 10^{15} GeV would be a hundred times bigger than the Solar System.) But there is a point which opens the theory to practical enquiry: the GUTs make a dramatic, unexpected, and testable prediction, namely, that the proton itself must be unstable and eventually decay radioactively. The very forces that brought us into being also ensure our eventual destruction. There is, however, no cause for immediate alarm because the lifetime of the proton predicted by GUTs is about 10^{32} years, almost infinitely large when compared with the present age of the Universe, about 10^{10} years. The current experimental lower limit on the proton's lifetime is already about 10^{30} years, however the GUTs prediction is accessible to test and although the experiment is very difficult several searches are in preparation. Paradoxically, an experiment telling us something about a time scale of 10^{32} years will give us confidence in talking about events on a time scale of 10^{-35} seconds. For those concerned about a life-span of only 10^{32} years I must add that more elaborate GUTs can, in fact, be constructed in which the proton is stable and that there are others who believe that even with decaying protons the rate of that decay decreases as the Universe ages so that only a minute fraction of the total would ever decay.

So by about 10^{-35} seconds we had the progenitor particles, the nature and variety of which depend on your choice among the GUTs, in the mass-energy range of 10^{15} GeV. The radioactive decay of the progenitor particles gave rise to the internal constituent particles of the hadrons, the quarks. As the Universe expanded particle–antiparticle annihilation took place, but a minute excess of particles survived, and eventually the density was sufficiently low for the quarks to cluster into the familiar hadrons: neutrons and protons and other particles that we know within the laboratory. This happened after about 10^{-6} seconds. As time passed the more highly excited of the product hadrons quickly disappeared and after about 10^{-5} seconds only neutrons and protons and a few of the lightest of the other hadrons, together with leptons of various kinds and, of course, photons were left. Beyond this time the evolution of the Universe depended chiefly on the number of distinct species of lepton that exist and related matters. Conventional nuclear and particle physics, accessible to laboratory investigation, had already taken over, the time scale stretches enormously and after about 3 minutes the course was set; most of the Universe's helium together with some other light nuclear species had already been manufactured; the rest is history.

But where are we today? What is the Universe like now and what will happen to it in the future? The Universe is made of stars, largely clustered into galaxies of various kinds, many of the galaxies flat, disc-like, and roughly spiral-shaped such as our own Milky Way; then there are odd things such as quasars that emit tremendous amounts of energy in a manner as yet by no means understood; probably some, perhaps many, black holes, a lot of gas, a lot of dust, and so on. But the whole thing is expanding as though everything were still rushing apart following that Big Bang. Will it go on expanding forever (although perhaps rather radically transformed after 10^{32} years or so by the GUTs decay of the protons and neutrons)? It is not certain that the Universe will expand forever, because gravity is obviously acting to slow down the expansion, to stop and reverse the motions, and bring everything back to where it started from; a Big Crunch followed perhaps by another Big Bang, or a Big Bounce. It is a question of how much mass there is in the Universe; if there is more than a certain mass then gravity will win in the end and everything will collapse into a Big Crunch—the Universe, in the jargon, is closed; if there is less than this mass then the expansion will continue forever—the Universe is open. Bookkeeping on the mass is difficult because of the amounts of gas and dust that lie, in quantities difficult to determine with precision, within and between the galaxies, because of the unknown number and mass of black holes, and so on. Indeed, there are within astronomy, at one or two places, what the astronomers call 'missing mass' problems. If you look at the motions of stars within galaxies you should be able to explain the dynamics in terms of the mass of the galaxy and the distribution of that mass within it, but when this is tried it does not come out right: it seems that there must be more mass within the galaxy than has so far been detected. Similarly when you look at clusters of galaxies and analyse the motions of the galaxies within the clusters, mass again seems to be missing. But even taking these problems into account present best estimates suggest strongly that there is not nearly enough mass there to close the Universe.

The open or closed Universe is not as yet, however, an open and closed book; there is a major point of reservation to be made. I have spoken of the weak force as transforming, for example, a neutron into a proton plus an electron, but this is not all; a third particle is emitted in that radioactive decay, an electrically-neutral lepton, the neutrino. The neutrino, whose interaction with both electrons and hadrons has been investigated experimentally in great detail, may be massless and has long been popularly supposed to be so. Now enormous numbers of neutrinos must abound in the Universe (the scenario we have followed implies that they are roughly equally as numerous as photons and so about 10^9 times as numerous as neutrons and protons). If neutrinos are massless they make no contribution to the mass of the Universe and are irrelevant to the question of the open or closed nature of the Universe. But they may have a mass and, indeed, very recent experiments suggest that the mass of the neutrino accompanying the beta-decay of the neutron is finite and perhaps as great as a few thousandths of a per cent of that of the electron. In that case the total mass of the neutrinos in the Universe would be so great as vastly to outweigh (literally) all other known matter, and could be sufficient to close the Universe, which would then collapse back into the Big Crunch in perhaps another 10^{10} years or so. No sooner are we relieved to learn that G U Ts proton decay nevertheless gives us another 10^{32} years than we find that we may be cut off after only 10^{10} years!

Before leaving the fate of the Universe it should be remarked that Einstein's equations of general relativity, on which the gross time-evolution of the Universe is thought to depend, admit of the addition of a term, the so-called cosmological constant, which could have the effect of forcing an expansion on the Universe even if the mass of the Universe were such as would, in the absence of that constant, close it. This cosmological constant was originally introduced when it was mistakenly thought to be necessary to secure acceptable solutions to Einstein's equations, but when the mistake was discovered it was seen to be unnecessary and so fell under the slash of Ockham's razor. But it could still be there and could, despite a finite mass for the neutrino, save us from the Big Crunch.

The Universe now

So how big, irrespective of its ultimate fate, is the Universe now? One can feel one's way out to this, starting from things with which one is familiar. First of all let us try to get a feeling for the size of the Earth. The curvature of the Earth is appreciable, although not usually perceived over short distances; if the ground were smooth and free of obstructions, if you went out of Wolfson College and, with your eye at ground level, looked down the road the 1.5 miles to Christ Church, the curvature of the Earth would cause the Dean to appear to be buried to his knees. You have a good grasp of 1.5 miles from ordinary personal experience, you have a good grasp of one-third of a Dean, you can begin to *feel* the curvature of the Earth. We could extend this picture to visualize the top of Salisbury cathedral's spire, 50 miles away and 1200 feet below the horizon; and so on. But today we are used to a view from the Moon: at a distance of 240 000 miles we see the Earth as a small globe, 8000 miles in diameter. Nearly 400 times further from the Earth, at 93 million miles, we find the Sun; this is a big step and now we should change to a more appropriate measure for big distances: the time taken for light to cross them. At 186 000 miles a second it takes Moonlight a little over one

FIG. 1.1. The President of Wolfson observes the Dean of Christ Church.

FIG. 1.2. The Earth seen from the Moon (NASA).

second to reach us, and Sunlight about 8 minutes. The nearest star to the Sun is α-Centauri which is about 4 years away, but there is another way of getting a feel for that if you have a feel for the distance of the Earth from the Sun and therefore the diameter of the orbit on which the Earth circles the Sun. Imagine, as the Earth goes around the Sun, that you fix your eyes on α-Centauri: because the Earth is moving around the Sun, α-Centauri will

FIG. 1.3. Andromeda. At a distance of about one million light years, this is the nearest galaxy to our own. About 10 000 times further we reach the limit of the visible Universe. (Hale Observatories.)

seem to move to and fro relative to the more distant stars that form the back-drop. The apparent motion of α-Centauri is about a second of arc and that is about the angle of a thumbnail at 3 miles, which you can picture all right. Beyond the nearest star distances begin to get quite appreciable and the effort to seize them also gets quite considerable. Our galaxy, the Milky Way, containing a few thousand million stars, is about 100 000 years across (although much thinner, being a spiral pancake); our nearest sister galaxy, Andromeda, is about a million years away but that is only a few hundred thousand times the distance of α-Centauri that you pictured through your thumbnail. Beyond Andromeda we had better go at once to the remotest depths of observable space, only about ten thousand times as far as Andromeda, and that is the end as far as we can see. The Universe is huge, without doubt, but you can almost grasp it, provided you can imagine the Dean of Christ Church sunk to his knees.

Nor is the observable Universe all that big in relation to the things it contains. The largest apparently structured objects that we know are observed not in visible light but at radio wavelengths through the techniques of radio-astronomy; they look rather like 'bow ties' and their appearance is due to electromagnetic rather than to gravitational forces. These objects are

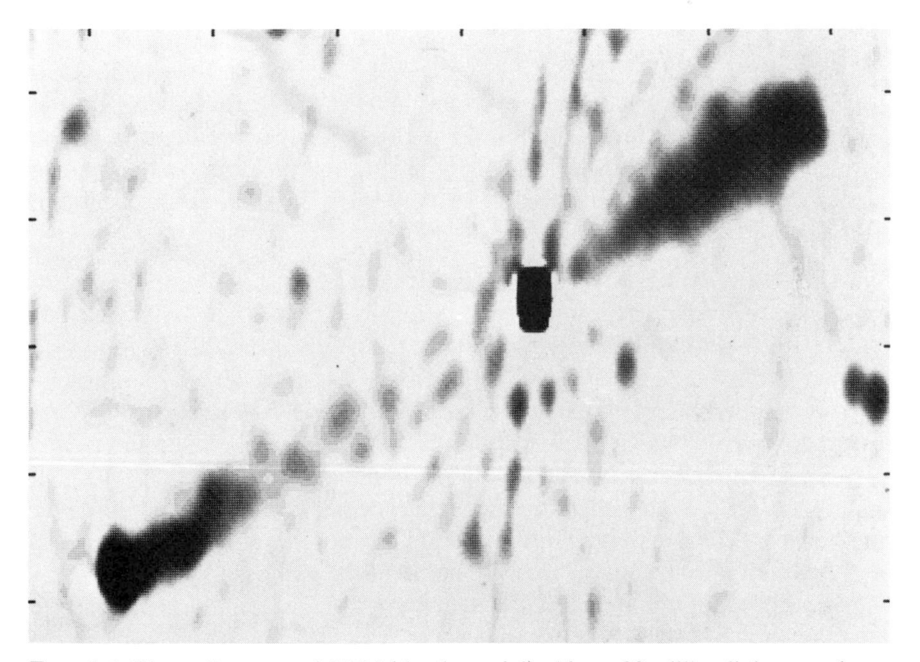

FIG. 1.4. The radio source 3C 236 (the 'bow-tie'). About 20 million light-years long, this is the largest structure so far found in the Universe. Although nearly 2000 million light years away, if visible to the naked eye it would appear somewhat larger than the Moon. In this photographic representation of the radio emission, observed by the Westerbork Synthesis Radio Telescope at 49 cm wavelength, increasing blackness represents increasing intensity of emission.

ordered streams of electrically charged particles spanning distances of some tens of millions of light years and so they stand to the entire Universe, as we can see it, just about as does a bow tie on the neck of a lecturer to a large lecture theatre. Large as the Universe may be, the forces that shape it manifestly reach out over distances not incommensurable with the totality of its size: imagine a lecture theatre with a large number of bow ties within it—we would scarcely say that it was empty.

Atoms

I have tried to persuade you to accompany me to the edge of the Universe, to persuade you at each stage that your imagination could take the next step. Let us now reach inwards towards and into the atom, to scales of distance vastly less than our own personal dimensions just as those of the Universe are vastly greater. Indeed, in discussing the earliest instants after the Creation we have already assumed there is sense in the discussion of objects that are smaller than ourselves by factors much larger than that by which the visible Universe is greater. What right have we to suppose that in either of these realms, so much larger and so much smaller than that of which we have direct personal experience, the language that we have devised to describe our personal experiences of events more or less of our own scale will continue to work? Lucretius had the same problem:

> The poverty of our language and the novelty of the theme compel me often to coin new words. This it is that leads me to stay awake through the quiet of the night studying how by choice of words I can display before your mind a clear light by which you can gaze into the heart of hidden things.

Similarly, how can we be sure that the mathematics that we have devised to codify our everyday experience will continue to work under conditions remote from that experience? May it not be that on scales of time and distance vastly beyond the realms of our personal experience, phenomena will obtain that are simply not susceptible of description through any thought process of which we are capable? That is to say, in our human terms, may there not be essential irrationalities in the very big and the very small such that we can never grasp their workings because to us they could never 'make sense'? This is an open question and I would beg you to leave it open, at the same time recognizing that the name of the game is to try to describe what we cannot possibly reach as though it obeyed the same rules as what we can. Recognize, as a corollary, that we may not be describing and understanding the Universe, very big and very small, but rather inventing it.

So now down to the atom. It is not very far away. We can easily see, through an ordinary microscope, things as small as 10^{-4} cm across—about a millionth of our own size. The atom is only about ten thousand times smaller and nowadays we can see that as well, not by bouncing light off it as one sees something with one's own eye but by bouncing high-energy electrons off it and then focusing them through electric lenses much as light is focused through glass lenses. What we then see (with the naked eye) on a TV-type screen exposed to focused electrons that have been scattered off a thin slice of crystal is a regular array of dots, each due to the scattering of the electrons from an ordered line of atoms in the crystal just as chemists and X-ray crystallographers have, for decades, been telling us we must. There are no surprises but it is nice to see the atoms properly regimented in their ordained rows. We can similarly 'see' single isolated atoms of heavy elements such as uranium scattered randomly over some light supporting surface and, more impressively, we can 'see' single molecules that contain more than one

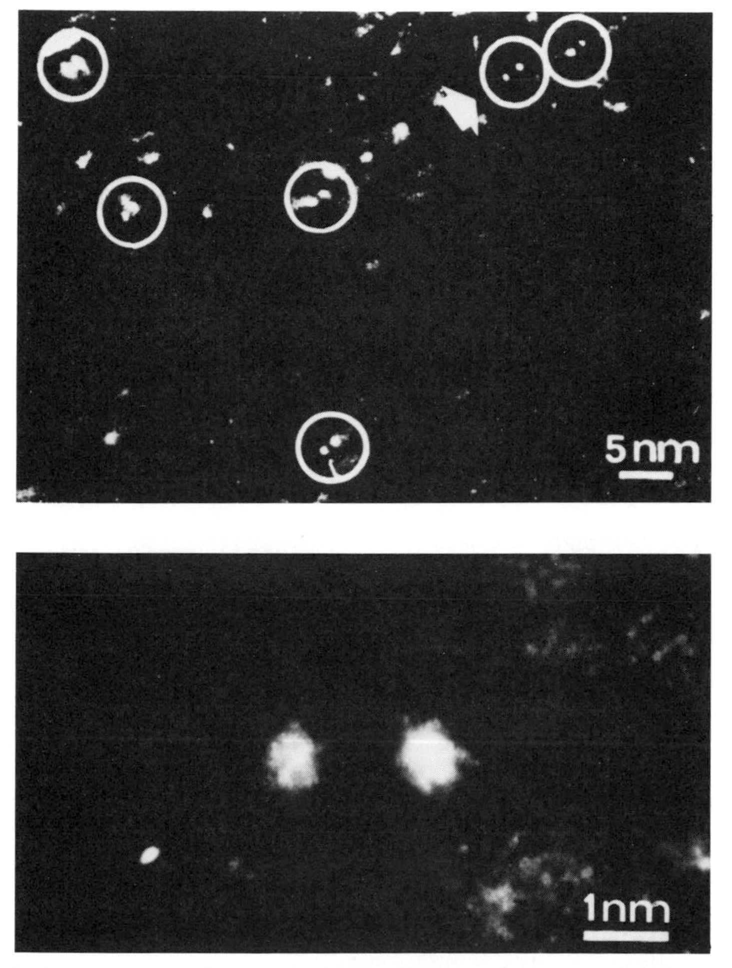

FIG. 1.5. Atoms can now be 'seen' with an electron microscope; the pictures show pairs of barium atoms, one at each end of an organometallic molecule only 1.6 nanometres (nm) long. (1 nm is 10^{-9} m, i.e. 10^{-7} cm or 10 Å). (Jouffrey and Dorignac, Laboratoire d'Optique Electronique, CNRS, Toulouse.)

heavy atom, the separation between the heavy atoms tallying exactly with our earlier inferential understanding. With electron beams we can also, by more indirect means analogous to the now-familiar optical holograms, go inside individual atoms and map out the distribution of their electrons in the 'line of sight' at various distances from the centre of the atom, as if you had plunged a minute cork borer through the atom and counted how many electrons you had bored out with it. Here also the results confirm what we had known for a long time, including such superficially surprising things as that the light atom neon has more electrons in certain regions of its anatomy than the heavier atom argon.

Atoms are like little Solar Systems somewhat more than 10^{-8} cm (1 Å) across, with electrons circling the central nucleus, itself only about 10^{-12} cm in diameter, made of neutrons and protons; the number of protons is the atomic number and determines the chemistry of the atom; the nucleus carries all but a few thousandths of a per cent of the mass of the atom. We should not say that the electrons 'circle' the nucleus, because quantum mechanics, the ground rules for talking about very small objects, does not permit the unqualified use of household concepts for discussing the very small, although what it requires us to use in their place are rather mixtures of household concepts; it does not require us to abandon everyday speech in our descriptions and certainly does not force us towards the irrationalities about which I speculated on switching from the very big towards the very small. In place of 'circling' as a planet does the Sun, the electron simply has its being within the atom, has a certain precisely defined probability, or chance, but only a *probability*, of being found at any spot if we suddenly dive into the atom at any arbitrary time. The electron also has associated with it a certain, also precisely defined, angular momentum of its motion around the nucleus of the atom. Quantum mechanics forbids us actually to watch the electron going round and round, or rather tells us that if we try to watch it we shall so gravely disturb its motion that the results of the observation will not be interpretable in terms of what would have been happening had we not looked; so there is no point in talking about 'what would have been happening' at all, since we can never check on it. Of course, planets going round the Sun should also be described quantum mechanically because quantum mechanics is a more fundamental law than classical mechanics, but we observe planets perfectly happily and do not fear knocking them off their orbits by so doing. There is no contradiction here: quantum mechanics tells us that the bigger a system becomes the more nearly do the laws of quantum mechanics approximate to those of classical mechanics and, indeed, even for an atom it becomes permissible to talk of an electron as circling its central nucleus if the atom is sufficiently 'excited' to move the electron out into an orbit of sufficiently large angular momentum.

How do we know that atoms are little Solar Systems (using that term within the above quantum-mechanical stricture)? Apart from 'seeing' them with the electron microscope, we calculate their properties, using quantum mechanics, and then carry out measurements to check those properties. In particular we can measure how much energy we have to put into an atom to remove an electron—that electron's separation energy—and compare that with calculation. When this is done we find very striking regularities: as we go to heavier and heavier atoms the amount of energy needed to prize loose an electron steadily increases, as we might expect on account of the increasing electrical attraction of the increasing positive charge on the central nucleus, but every now and again there is a sudden drop in the energy needed, followed by a subsequent rise from the new low. This is precisely what is predicted by the quasi-Solar-System model coupled with an extraordinary quantum mechanical rule called the Pauli exclusion principle, which limits, rigorously, the number of electrons permitted to occupy one 'planetary orbit' and demands that the next go into a more remote, and therefore less strongly bound, orbit.

So we have very good confidence when we speak of atoms and of their building up into molecules of varying degrees of complexity, from the simple H_2O to the giants that are associated with the processes of life. However the statement that an atom is only ten thousand times smaller than something that you can actually see with an ordinary microscope may perhaps give the impression that atoms are large, ungainly galumphing things; so they are, compared with what we shall look at next, but they are, nevertheless, quite small and very numerous. Classical illustrations of the abundance of atoms and molecules have always been popular: that your next cup of tea will contain about a thousand molecules of water from the cup of hemlock that Socrates was forced to drink, assuming that Socrates' cup has by now been fully mixed into the oceans and atmosphere of the world; that, at this moment, the lungs of each one of you contain about one molecule from the expiring breath of Julius Caesar (and, for that matter, of Genghis Khan and Helen of Troy . . .), again assuming a uniform mixing of expiring breaths throughout the atmosphere. But there is no need to over-sensationalize a theme, which is surely dramatic enough in itself.

Now to the central nucleus, the Sun of the atom. How do we know that it is there and how big is it? If one fires accelerated atoms into a gas, or a thin foil of matter, they will be deflected from a straight path, or scattered, because they contain electrical charges and so do the atoms on which they impinge. (Although individual atoms are electrically neutral, electrons can be got out of them, as we have just noted, and so the overall electrical neutrality must be made up of some sort of balance between negative and positive constituent charges.) We can measure the probability that the impinging atom is scattered through very small angles and then can confidently predict the probability that the impinging atom will be scattered through a large angle *if* that is just due to the chance accumulation, as it passes through or near many atoms of the target gas or foil, of many small nudges in the same direction. The result is sensational: the probability of scattering through a large angle is totally inconsistent with expectation derived from the scattering through small angles. In the experiment carried out by Rutherford in 1911 swift helium atoms impinged on a thin gold foil and the probability of scattering through more than 90° was about $10^{10\,000}$ times greater than expected from scattering through small angles. It is impossible to imagine so large a number; it is much easier to hold the Universe in the palm of your hand. Rutherford said that he was as astonished as if a 16-inch shell had bounced backwards from a sheet of tissue paper. Obviously atoms are something other than just vague balls of mixed positive and negative electricity. Rutherford's quasi-Solar-System model to explain this tremendous large-angle scattering was that at the centre of each atom is a 'point-like' nucleus carrying as much positive charge as all the circling electrons carry negative charge and also carrying almost all the mass of the atom. When, now, two atoms impinged 'head on' the two nuclei approached each other and, with the entire positive charge within each atom being concentrated at a point rather than spread out through the entire volume of the atom, the nuclei could exert vastly greater forces upon each other, as they came close together, than if the charges had been spread out. The spread-out model was a good enough approximation to describe the small angle scattering when the nuclei remained far apart, but not when they came into close collision. It is easy to work out the law of scattering probability as a function of angle of scattering for this nuclear model and it fitted perfectly: the discrepancy of $10^{10\,000}$ disappeared.

Speaking of the positive charge of the nucleus we used the phrase 'concentrated at a point'; this, of course, is not so but it works as though it were true so long as the nuclei do not touch. If the nuclei do touch then the simple point-scattering law breaks down and from the conditions under which this happens we can infer the size of the nucleus: about the 10^{-12} cm mentioned earlier. Since, so far as we know, electrons are point charges and since the atom is about 10^{-8} cm in its own dimensions the volume occupied by the nucleus is only about 10^{-12} of that of the atom: the atom is almost entirely empty space.

I have several times mentioned neutrons and protons: they are the constituent particles of the nucleus, almost identical to each other in mass and other properties apart from their electrical charge which for the proton is plus one (the electron being minus one) and for the neutron zero. The neutron and proton themselves have size—about 10^{-13} cm—as we can discover by bouncing them off each other or by bouncing electrons off them. When the electron penetrates inside a proton it feels a smaller attraction than if the proton had been a point because part of the proton's charge is then trying to pull the electron out and only the remainder to pull it in, so the resultant scattering is through a smaller angle than it would have been for a point proton; in this way the proton's (electrical) size can be measured and, indeed, the distribution of its charge effectively mapped. The neutron, although electrically neutral over-all, also scatters electrons because it has an internal electrical structure that, as for the proton, can be mapped by the scattering.

The neutrons and protons inside the nucleus are not packed tightly together and indeed they move quite freely around in almost independent orbits like the electrons far away in the outer space of the atom's Solar System. We have, in effect, a quasi-atomic model for the nucleus itself, the nuclear shell model; there is no special centre to the nucleus, it has a uniform density and the binding force experienced by any neutron or proton is just the sum of the forces on it due to all the other neutrons and protons of the nucleus. The shell model shows up phenomenologically in much the same way that the electronic orbital model of the atom did: by systematic changes, with increasing mass of the nucleus, of the neutron and proton separation energies and with sharp drops in those energies when the next neutron or proton is forced, by the Pauli exclusion principle, into a higher orbit.

There is a tremendous wealth of nuclear phenomena; we know and have studied the properties of many hundreds of nuclear species, each being characterized by its own number of neutrons and protons. By-and-large we understand what is going on in terms of many collective phenomena which are all entirely consistent with the quasi-atomic shell model and which, indeed, can be derived from it but which form more succinct and appealing ways of describing certain of the modes of organization of the neutrons and protons: we speak of the nucleus as being deformed into ellipsoidal or hour-glass shapes, of rotating or of vibrating, with all the possible combinations of these gymnastics; we speak of oscillations in which all the protons and all the neutrons swing backwards and forwards relative to each other in two inter-penetrating sets; and so on, but none of these ways of speaking is other than a convenient short-hand to summarize what would be a somewhat elaborate description in terms of the quasi-atomic shell model.

Here we shall leave the rich variety of nuclear phenomena; they still pose many interesting questions but in the present context of an attempt to paint, with an exceedingly broad brush, the whole range of our vision of the organization of the Universe, we can consider nuclei to be fairly well understood.

Messengers of force

Before moving to sub-nuclear matters we must go right back to the beginning when I spoke of the four forces of Nature. How do forces come about? What makes the inverse square law of the electric force and the more sharply cut-off law of the strong force, which drops very quickly to zero after about 10^{-13} cm? There is, among physicists, an abhorrence of 'action at a distance'. Crudely speaking we have a revulsion from any theory that speaks of an interaction—a force—between particle A and particle B without saying how particle A becomes aware of particle B's presence. In other words we demand (admittedly on philosophical, or possibly even sentimental, grounds) that interaction should depend on communication: A and B cannot know, cannot experience a mutual force, attractive or repulsive, unless they find out, via an appropriate messenger, about each other's existence. The messenger, of course, can only be some other particle which A emits and tosses across to B and vice versa.

We therefore believe that all forces in Nature (*pace*, for the moment, gravity which remains somewhat enigmatic) are generated by the exchange, between the interacting particles, of messenger particles. Let me give a homely illustration of a force generated by a messenger particle. You are observing, from afar, two skaters on a frozen lake: they are drifting along

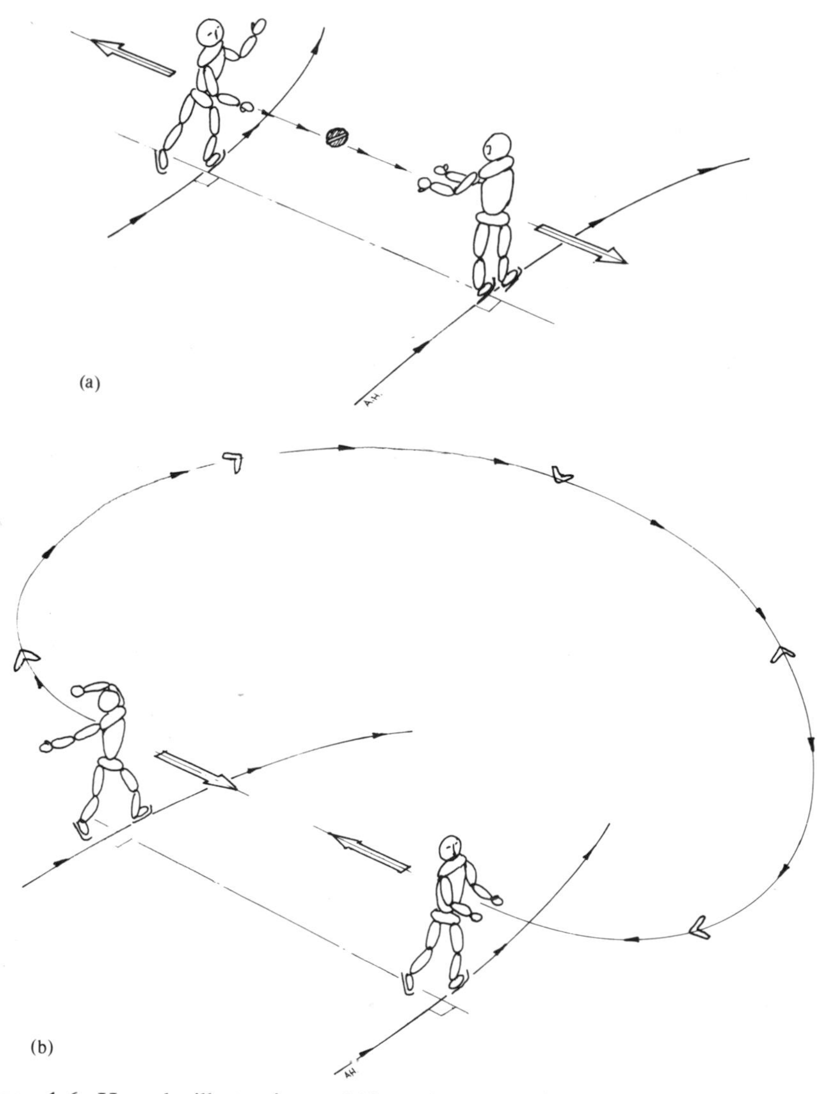

FIG. 1.6. Homely illustrations of 'forces' generated by exchange: (a) repulsive skaters exchange a cricket ball; (b) attractive skaters exchange a boomerang.

side by side, a few feet apart, but then, without apparent movement of their legs, begin to move further apart: a repulsive force appears to exist between them. You put up your binoculars and observe that although they are doing nothing else they are throwing a cricket ball to and fro. Skater A throws the ball to skater B; A recoils from his throw, moving away from B and B recoils from his catch, moving further away from A; and so on and so on. The cricket ball is the messenger particle that has established a repulsive force between A and B. You may accept this home-spun illustration of the establishment of a *repulsive* force due to exchange, but how about an attractive force? Consider the same example of skaters on a lake but imagine that, from afar, you see them coming closer together: the force between them is now attractive. Put up the binoculars again: this time they are exchanging a *boomerang*. Skater A throws the boomerang away from skater B and A therefore recoils *towards* B. The boomerang curves through the air away from him and then towards B, passes him and then curves back from the other side so that B catches it when it is moving towards A and therefore gives B a push *towards* A, the force is attractive; and so on and so on. (There will be those among my readers who will consider, and quite rightly, that this illustration is skating upon thin ice.)

What are these messenger particles? In the case of the electromagnetic force they are the photons, the quanta of electromagnetic energy. The nuclear force can be communicated by a whole host of messenger particles called mesons. Now the reach of a force, the extent of the space over which its influence can be felt, depends on the mass of the messenger: the heavier it is the shorter the range. Thus the photon which has zero mass (although it does carry energy) may, technically speaking, extend the influence of the electromagnetic force to infinity. (The famous inverse square law arises simply because the area of a sphere of radius r increases as r^2 so that the chance that a photon despatched randomly by A will light upon B at a distance r goes as $1/r^2$.) The lightest meson is the pion with a mass about 275 times that of the electron and this explains why the strong, nuclear force drops off so sharply beyond about 10^{-13} cm. There are many other mesons of greater mass that contribute to the strong force at smaller distances and which give that force a rather complicated form. The weak force we believe to be transmitted by particles, also mesons, called intermediate vector bosons, $W^{\pm}$ and Z^0, of mass about 100 times that of the proton and hence limiting the reach of the force to a few times 10^{-16} cm. Thus we picture the beta-decay of a neutron into a proton plus an electron plus an antineutrino as proceeding by the emission of a W^- particle from the neutron, which latter thereby becomes a proton, followed 10^{-26} seconds or so later by the disintegration of the W^- into the electron and the antineutrino. These $W^{\pm}$ and Z^0 particles are intimately linked to the photon and specify the energy scale relevant for the electro-weak unification of the electromagnetic and weak interactions. The possible GUTs unification energy of the electroweak and the strong interactions is also associated with a messenger particle, named *lepto-quark*, whose mass would be of the order of the unification energy of about 10^{15} GeV; this same particle is the progenitor, the one with such a crucial role in the birth of matter near the beginning of this Universe.

Proton structure

We now turn to the proton and neutron themselves. They used to be called 'elementary particles' but that name is clearly no longer appropriate if by it you mean a thing entire unto itself, having no constituent parts or wheels going round inside it. That protons, in fact, have a structure can be inferred in several ways. Perhaps the most appealing is the direct analogue of Rutherford's experiment on the scattering of swift atoms by matter that revealed the nuclear structure of the atom so dramatically, showing unequivocally that atoms have little hard bits inside them. When protons bombard each other at very high energies they tend to emit mesons, as is only to be expected since it is by the emission and absorption of mesons that protons communicate with each other. If protons are simply pieces of matter of some sort, with extension but no internal structure, then, just as in the case of the highly-probable small-angle scattering of Rutherford's experiment, we can observe the copious emission of mesons at small angles relative to the collision axis of the protons, emission corresponding to the relatively gently nudging of the protons by each other, and confidently extrapolate the production to large angles of meson emission. More accurately, we extrapolate the production of mesons with a small transfer of transverse, sideways, momentum between the participants to the production of mesons of high transverse momentum transfer. When we do this we get the same shock as Rutherford: at high momentum transfer the production of mesons is many orders of magnitude greater than we expect from our extrapolation of the low momentum transfer production. Just as surely as there are little hard bits inside atoms there are little hard bits inside protons to give that unexpectedly frequent high transverse momentum transfer; protons and neutrons and, we presume, all hadrons are structured objects containing constituent particles much smaller than themselves.

The best limits on the size of these constituent particles come from bombarding protons with very energetic neutrinos, point-like particles interacting only via the weak force. Let us assume, for the moment, that the weak interaction between the neutrinos and any proton constituents is of effectively zero range. Then, for reasons to do with basic quantum mechanics, if the constituents within the proton can be regarded as point particles the probability that a neutrino, on impinging upon a proton, causes it to emit another particle—a lepton of some sort—will be proportional to the energy of the neutrino. But if the constituent sub-structure of the proton, with which the neutrino is interacting, itself has finite size then this simple proportionality will break down; the energy of neutrino at which this happens is a measure of the size of the constituent particle. So far, in experiments involving neutrinos of energies up to 100 GeV, this proportionality holds well enough for us to state both that there are 'little hard bits' inside the proton no bigger than, at most, 10^{-15} cm and that any finite range of the weak force is also less than this. (Indeed, the predictions of the electro-weak theory suggest the range is about 10^{-16} cm.) Since the proton itself is about 10^{-13} cm across this small limit on the constituent size suggests that protons and neutrons themselves, like the atom, are chiefly empty space. These remarkable experiments with neutrinos are described by Donald Perkins in chapter 4. They also enable us to count the number of 'little hard bits' in the proton: it comes out at 3 or close to it.

Another, totally different, line of argument also suggests strongly to us that the protons, and all hadrons, are indeed made out of some sort of constituent particles. If we take any structured system: a molecule made out of constituent atoms, an atom made out of constituent nucleus and electrons, a nucleus made out of constituent neutrons and protons, we can rearrange the constituents of that system relative to each other and thereby raise that system into states of higher energy, called excited states. We can make the atoms in the molecule vibrate against each other, we can elevate the electrons of the atom into higher orbits, we can make the neutrons and protons of the nucleus vibrate or move to higher orbits; all structured systems exhibit a tremendous richness of excited states simply because their constituent sub-units can move relative to one another in such a variety of ways. Indeed, if one is shown a diagram (Fig. 1.7) of the excited states of one of these objects, molecule, atom, or nucleus, one cannot tell to which object the diagram refers without being told the energy scale: meV or eV for molecules, eV for atoms, keV or MeV for nuclei. Now if energy is given to a proton, say by bombarding it with another proton or by any other means, it can be raised into a whole host of excited states, just as can the other structured systems, the only difference being the energy scale: not now meV or eV or keV or MeV but rather tens or hundreds of MeV. The same is true of the mesons; they too can evidently be raised into excited states. There is no avoiding the conclusion that the proton, and all other hadrons are, like the nucleus, the atom, and the molecule, structured objects.

What are these constituent particles, the 'little hard bits' of the proton and all hadrons? To cut a long and intensely fascinating story very short: when one looks at the systematic relationships between the excited states of all the hadronic particles that are now known—the number of excited states running into hundreds thanks to enormously painstaking work over the past 30 years at the great accelerators in the USA, Europe, and the USSR—we find that it all makes sense in terms of the constituent particles that Murray Gell-Mann dubbed *quarks*. Hadrons such as the proton should contain 3 quarks (recall the experimental figure quoted above from the neutrino experiments) while the mesons should contain one quark and one antiquark. We know nothing to contradict this beautifully simple picture; the

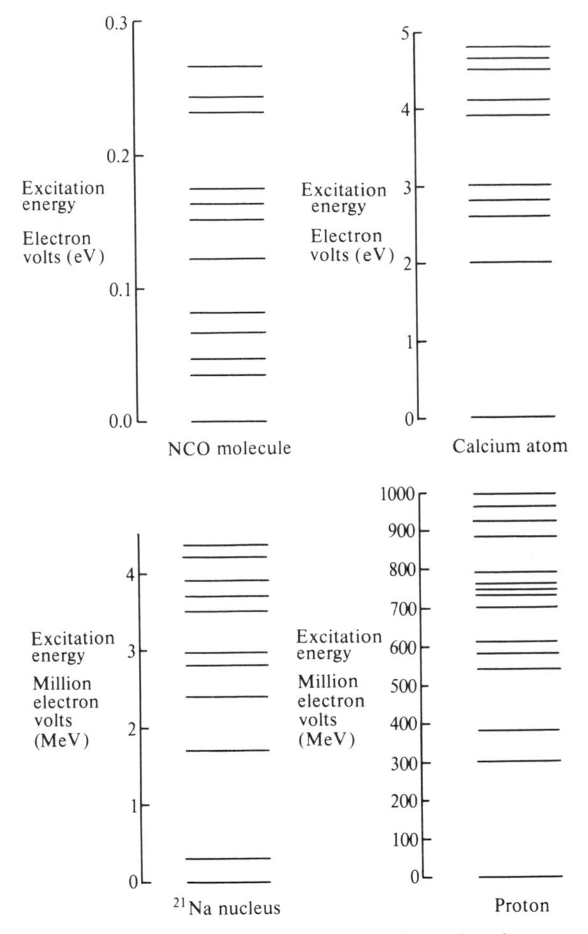

FIG. 1.7. Energy level diagrams for excited states of a molecule, an atom, a nucleus, and the proton.

excited states are simply rearrangements of the constituent quarks relative to each other like the rearrangements of the electrons that give rise to the excited states of their atom, or the atoms whose rearrangement gives rise to the excited states of their molecule, or the neutrons and protons whose rearrangement gives rise to the excited states of their nucleus.

There are several species or 'flavour' of quark; there are 'ordinary' quarks (of two kinds called 'up' and 'down') such as go into the neutron and proton and their excited states; there are 'strange' quarks that, by themselves or mixed with the 'ordinary' quarks, made the 'strange' particles that come out of energetic collisions at the great accelerators but that take so long over their subsequent radioactive decay that it is as astonishing as if Cleopatra had fallen off her barge on the Nile in 35 B.C. but splashed into the water only today; there are 'charmed' quarks that have other distinctive properties again and 'bottom' quarks and, probably, but not yet pinned down, 'top' quarks. And there, perhaps and perhaps not, the list will close. But all the myriads of hadronic particles that we know of can be constructed out of this short list of quarks and their antiquarks.

A very obvious question: what holds the quarks together inside the proton and the other hadrons? An associated question is why we have so far failed, despite the blandishments and importunings of our most powerful accelerators, to liberate individual quarks from hadrons no matter how energetically we have bashed hadrons together. Atoms can be got out of the molecules that they compose; electrons, part of the structure of atoms, can be liberated from atoms; nuclei can be dissociated into their constituent neutrons and protons; why, if quarks are inside neutrons and protons cannot they be got out of them? Lucretius would not have worried about this:

> There is an ultimate point in visible objects which represents the smallest thing that can be seen. So also there must be an ultimate point in objects that lie below the limit of perception by our senses. This point is without parts and is the smallest thing that can exist. It never has been and never will be able to exist by itself, but only as one primary part of something else. Since they cannot exist by themselves, they must needs stick together in a mass from which they cannot by any means be prized loose.

But today we must confess that we must find an answer to why we cannot prize loose something that we confidently assert to be inside. (Do not confuse this with the inverse problem: the fact that something comes out of something does not require that it should have been there in the first place; an electron and an antineutrino come out of a neutron leaving behind a proton, as we have noted, but they were not there to begin with; both were manufactured in the act of radioactive beta-decay of the neutron into a proton; barks come out of dogs but dogs are not made of barks. However, if we say that something *exists* inside something we have to be able to say why we cannot get it out.)

On the guiding principle that force is communicated by the exchange of a messenger particle between the interacting objects, the answer to the first question, namely what holds quarks together inside hadrons, is that *within* hadrons, acting between the quarks, is a particle, somewhat unimaginatively named the gluon, that sticks quarks together to form hadrons. The second question as to why we do not seem to be able to prize quarks out of hadrons is not yet satisfactorily answered but a partial answer is that when you look into the force that is generated between quarks by gluon exchange you find that, unlike the other forces we have encountered, it *increases* as the separation between the particles increases, rather than decreases; the closer the quarks, the *weaker* their interaction. (Surprising but not perhaps all that surprising—think of a rubber band.) This property, technically known as quark confinement, is associated with another property of quarks. It seems that each flavour of quark—up, down, strange, charmed, bottom, and (we anticipate) top—comes in three varieties called the 'colours'. Thus we speak of, say, red, yellow, and blue 'up' quarks and similarly (the same colour range) for the others. (It must surely go without saying that, in this context, colour is only a label, a name, and has nothing to do with the red, yellow, and blue of our own perceptions; we could use any other name and refer, for example, to Conservative quarks, Labour quarks, and Liberal quarks.) The reasons for this extra label of colour are compelling but rather technical and will be described in chapter 4 by Donald Perkins. It turns out that the gluons that flit between the coloured quarks must also carry colour and from this follows naturally, but by no means obviously, the property of confinement or at least an interaction that increases with distance of quark–quark separation until we meet the analogue of the snapping of a rubber band. But even this does not release the quarks; when the energy put into the system through the effort of separating the quarks against the force of confinement exceeds the energy needed to make mesons then one, or more, of these particles is created; we get new quark–antiquark pairs but no liberated quarks.

This quark–quark force is the strongest of any of which we are yet aware; the strongest force we know in Nature is that which operates not *between* neutrons and protons but *within* neutrons and protons. How is this related to what we earlier termed the 'strong' force, that which operates *between* neutrons and protons, holding them together inside the atomic nucleus? There is more than one way of answering this question but the best present answer is by analogy with the force that operates between atoms as they approach each other. Within each individual atom we have the inverse square law of electric force between the nucleus and the electrons and between electron and electron, which, following the rules of the quantum mechanical game, establishes the structure of the atom. But when one atom looks at another from a distance it is not, as we have already noted, electrically impressed, because the viewed atom appears to the viewer atom as electrically neutral. The summed electron charges of the 'orbiting' electrons are exactly balanced by the positive charges on the protons of the central nucleus; although the electric force is an inverse square force with distance it has totally cancelled itself out because of the over-all electrical neutrality of the whole atom. Indeed at sufficiently large distances of separation the extremely feeble gravitational force then wins out over the electrical force. But now, as the two atoms tentatively approach each other they begin slightly to interpenetrate; electrons 'belonging' to one atom may, if they stray far enough away from their parent nucleus, sense the presence of the other atom's electrons and nucleus. As they interpenetrate a force is established between them, exceedingly feeble at first but increasing very rapidly as the interpenetration increases. This atom–atom force is known as the 'van der Waals force' and, empirically, looks very different from the inverse square electrostatic force; it increases, indeed, something like the inverse seventh power of the distance between the atoms. However it is in no fundamental sense a force in its own right but only a representation of a particular aspect of the inverse square law of electrostatic force operating within the laws of quantum mechanics between extended and structured electrostatic systems that each possess over-all electrical neutrality. In the parallel situation of the quark–quark force due to gluons and the proton–proton force, protons, remote from each other, have no strong interaction to speak of (their electrical interaction is irrelevant here: if you wish, think of the interaction between two neutrons) because the gluons are fully occupied with the intra proton forces that hold the proton's constituent quarks together. But as the protons approach each other, the quarks in the one proton sense the presence of the quarks in the other. How do they do this? Is it by the exchange, between the quarks of the one proton and the quarks of the other, of the same gluons that give rise to the quark–quark force within each proton? This is tricky to arrange in terms of the book-keeping of colour but it might be done. Is it by the actual percolation of the quarks themselves between the protons? Is it by the creation, through the gluon force, within one proton, of a fresh quark–antiquark pair which then, as a 'conventional' meson, trips to the other proton? The answer is probably complex: at large distances of separation the last possibility is probably right and at short distances the other two mechanisms must certainly come into play. But however we look at it, the strong force between two protons is due to a sort of leakage, from one proton to the other, of the basic strong gluon-exchange force that operates *inside* each proton between its constituent quarks. In this, qualitatively clear, sense the proton–proton strong force stands to the fundamental quark–quark strong force much as the atom–atom van der Waals force stands to the basic electrostatic force within each atom.

Epilogue

That is about as far, but not quite as far, as I should wish to go in this stage-setting introductory chapter. There is one outward-looking and one inward-looking final remark that I should wish to make. The outward-looking remark: What is so special about us and our Universe? We do not know that there are not many more Universes, perhaps constituted of different sorts of matter and containing different sorts of consciousness from our own, perhaps even—most excitingly—capable of exchanging exceedingly weak but detectable signals with our own; the anthropic principle, that says that our Universe has to be as it is or we should not be here to know that it is as it is, is a powerful one but also tells us that there are indeed other Universes elsewhere and elsewhen, or perhaps, as I say, here and now. The inward-looking final remark: what about 'lesser fleas', are the quarks the end or do they not have their own sub-structures? Look into a bacterium and you find it is made of molecules; look into a molecule and you find it is made of atoms; look into an atom and you find it is made of electrons and the nucleus; look into a nucleus and you find it is made of neutrons and protons; look into a neutron or proton and you find it is made of quarks; look into a quark and . . .? I am sure that I need not tell you that speculation is rife. The only argument against such infinite regression is that of Lucretius:

> (in that case), however endlessly infinite the Universe may be, yet the smallest things will equally consist of an infinite number of parts. Since true reason cries out against this and denies that the mind can believe it, you must needs give in and admit that there are least parts which themselves are partless.

Perhaps we should remind Lucretius that: '. . . no fact is so simple that it is not harder to believe that to doubt at the first presentation.'

2

Particles and Forces

SIR RUDOLF PEIERLS

The world around us consists of particles held together by forces. Of these two concepts the forces were originally the more familiar; the longest-known, and the most familiar one, is the force of gravity. It was described in detail by Isaac Newton, who formulated clearly not only the law of gravity, but with it also the laws of mechanics, which describe how bodies move under the influence of forces, such as the force of gravity. In doing so he discovered something that was very surprising: he found that the laws he had formulated applied equally to the very large and the very small. The laws of nature seemed to know no natural scale. That was by no means an obvious matter. To our intuition the large and the small appear very different, and it came as a revelation that the same laws would serve to describe both the motion of the Earth and the planets around the Sun, and that of the wheels inside a watch. The success of Newton's mechanics, with its scale-independent laws, was immediate, and from then on it was taken for granted that natural laws must always be scale-independent.

Another result of Newton's success was that it now seemed that mechanics, the science of force and motion, must be the basis of all physics, indeed of all natural science. To explain anything one had to specify a mechanism. This, originally, was supposed to apply also to the phenomena of electricity and magnetism. Static electricity had been known for long as an oddity, and magnetism as an equally mysterious force of nature, but rather more useful through the navigator's compass. Since then, electromagnetism has come to play a much more important part in our lives. We have become accustomed to the availability of electric power from outlets in our houses, without being conscious of its origin. Only the energy crisis of recent years has made us more aware of the problems involved in its generation.

On the face of it, the phenomenon of electric force, that is the mutual repulsion of like, and the attraction of unlike, charges appears to be an action at a distance. The presence of an electric charge somewhere in space would seem to cause directly a force on another charge some distance away. However, such action at a distance may only be apparent. If we find ourselves outside a door with an old-fashioned bell-handle to pull, it would appear that pulling the knob directly activates a bell some distance away in the house. But we know that, in fact, there is a wire running from the knob to the bell, and the tension caused by pulling the knob is transmitted from point to point along the wire, so that each piece of wire is affected only by the next piece in direct contact with it.

It seemed natural to suspect that electric and magnetic forces were similarly transmitted from point to point, and this idea led Faraday to consider *lines of force*. These are lines which run from a positive to a negative charge, or from a magnetic north to a south pole, in such a way that they are everywhere pointing in the direction of the force that would act on a charge, or a magnetic pole, placed at the point in question. Fig. 2.1(a) shows the familiar pattern of lines of force near two opposite charges or poles, and Fig. 2.1(b) shows the pattern obtained by spreading iron filings near two magnet poles, when the filings tend to take the direction of the local force and thus make the lines of force visible.

This idea of lines of force developed into the concepts of the electric and magnetic *fields* pervading all space. Finally James Clerk Maxwell formulated the equations determining the behaviour of these fields. At first it was believed that such electromagnetic actions had to be transmitted by some mechanical medium, called the 'ether', and even Maxwell at first regarded his equations as describing the state of tension or stress in the ether.

The reason for this was the prejudice that every physical phenomenon must have a mechanical explanation. It took a great effort, even for Maxwell, to rid himself of this prejudice, and to realize that the electromagnetic field was a basic concept of physics in its own right. Its laws could not be derived from those of mechanics; they did not contradict them but enriched the world of physics.

The great success of Maxwell was to unify our understanding of electricity and magnetism, to show that they were different aspects of the same phenomenon. He also showed that the transmission of action in the electromagnetic field, if the changes were rapid enough, could take the form of waves, rather as the wire in our bell pull, if jerked rapidly, would start vibrating. In other words, Maxwell's equations predicted the existence of electromagnetic waves, and thus explained the nature of light.

Electromagnetic waves comprise much besides light. The longest waves in practical use are radio waves; going down in wavelength we come to the range of waves used in television channels and in radar. Shorter still are the waves of infra-red or heat radiation which cause the pleasant sensation we experience in front of a fire. Visible light covers a limited range of wavelengths, and beyond it comes the ultra-violet range, similar to light, but invisible to the human eye, followed by X-rays, and shorter still the so-called

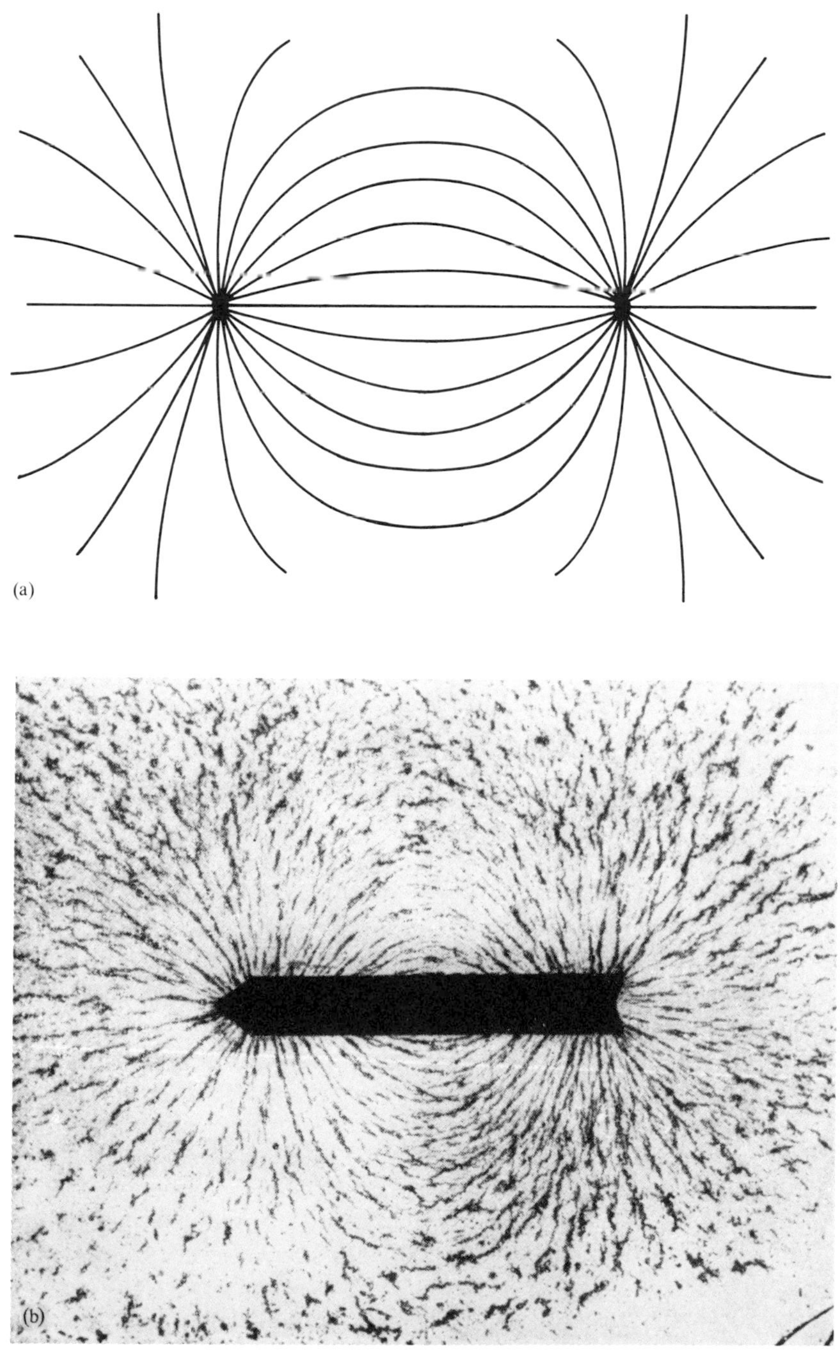

FIG. 2.1. (a) Lines of force in the vicinity of two opposite electric charges or magnetic poles. (b) Iron filings reveal the pattern of lines of force in the magnetic field surrounding a bar magnet.

γ-rays. Here we run out of names, and all waves beyond this range, however short, are still referred to as γ-rays.

In a sense, Maxwell's theory of electromagnetism and light completed the basic structure of physics as it was understood in his time, because it was only about the turn of the century that one came to consider the structure of matter a proper field of study for the physicist. Originally that domain belonged to the chemists, who had already shown that matter almost certainly consisted of atoms. The final confirmation of the atomic hypothesis required a means of determining the size of an atom, and this was done by physical methods.

The structure of atoms

As you know an atom is a very small object: if every inhabitant of the United Kingdom put down one atom, and these were placed in a row, the total length would be but a few millimetres, about one quarter of an inch.

Further insight came from the discovery of the *electron* by J. J. Thomson. The electron, evidently a constituent of the atom, is a very light particle, light even on the atomic scale, carrying a negative electric charge. The next step was Rutherford's discovery in 1911 that the atom consisted of a small positively charged centre, the so-called *nucleus* carrying almost all the atom's mass, with the electrons orbiting around it rather like planets around the Sun. Atoms of different chemical elements differ by the amount of charge on the nucleus, and hence by the number of electrons which will collect around that nucleus to make the whole atom electrically neutral.

This was an attractively simple picture, but there was a paradox. It could not be quite like the Solar System, because the scale-independence of the mechanical laws to which I have referred means that there is no prior reason for the size of the planetary orbits. The actual distance at which the Earth, for example, circles the Sun is determined purely by the historical accident of the way the Solar System developed. Yet all atoms of a given species, for example all hydrogen atoms, are alike, and if we eject the electron from a hydrogen atom and it eventually picks up a new electron, the atom's size, its energy, and all other properties will be the same as before.

The planetary orbits remain the same only because there are no disturbances to deflect them from their paths. Artificial satellites moving close to the Earth are disturbed by air resistance, which ultimately causes them to crash down. Atoms are subject to similar disturbances, because different atoms are close enough together, and also because an orbiting electric charge tends to produce electromagnetic waves which carry away energy. So we would expect the atomic electrons to crash into the nucleus, and this does not happen. Atoms are evidently stable, otherwise we would not be here.

The resolution of this paradox took a generation to sort out. On the small scale of the atom the laws of physics are no longer scale-independent, but new features appear. It is not possible within the scope of this chapter to give an exposition of the *quantum theory*, which embodies these new features, but I can stress one aspect of this: the motion of every particle, such as an electron, has associated with it a wave, of a wavelength depending on the mass of the particle and the speed with which it moves. The region in which the particle moves cannot shrink to a size less than this wavelength.

Thus, if we want to picture the hydrogen atom, which consists of just one electron attracted by the lightest nucleus, a *proton*, it should not be as a 'planetary' particle moving in an orbit around a nuclear 'sun', but rather like Fig. 2.2(a) which shows a sort of 'contour map' of the hydrogen atom. In the

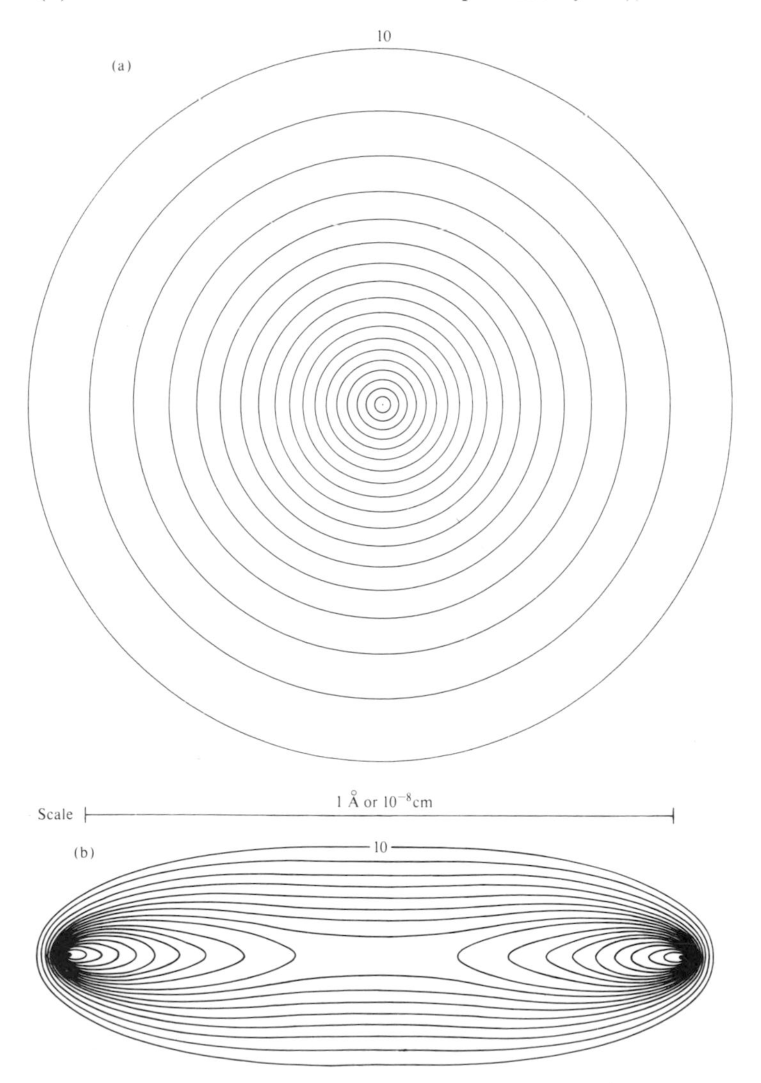

FIG. 2.2. (a) Hydrogen atom. (b) Hydrogen molecule ion (H_2^+: one electron and two protons). The 'highest contour lines' shown have a value 95, the others occur at steps of 5 units in relative probability to find the electron at a point.

normal state of the atom the most likely spot for the electron to be is at the centre; the likelihood falls off, rapidly at first but then more slowly, as we move radially outwards. There is a 10 to 1 chance against finding the electron anywhere farther from the proton than 1.4×10^{-8} cm, so we may regard this distance as the radius of the atom with the electron, most of the time, flitting rapidly about inside.

So on the small scale, when the wavelength is no longer negligible, the laws of mechanics have taken quite a new form, which we call quantum mechanics. Strictly speaking, it is not just a matter of smaller dimensions. The wavelength associated with the motion of a particle of mass m and speed v is given by h/mv, where h is Planck's constant and is a very small number (if m is in grams and v is in cm/s then $h = 6.6 \times 10^{-27}$). For example if we move an electron at a speed of 30 miles an hour, its wavelength is about one twentieth of a millimetre—about the thickness of a hair, a dimension quite visible to our eye. But at the speeds with which electrons move in atoms, about a million m.p.h., the wavelength is only about the size of an atom. There is an important corollary to this: the smaller the structure we wish to explore the smaller must be the wavelength of the particles we use and so the greater the value of their momentum, the quantity mv. For example, the electrons accelerated by strong electric fields in an electron microscope have a wavelength much shorter than light and enable us to study viruses and other structures too small to be seen with an ordinary microscope. To go further down the scale of dimensions requires still higher momenta, or energies, and today's limit is set by the size and cost of the huge high-energy particle-accelerators described by John Adams in chapter 7. It is quantum mechanics which dictates the need for high energies to probe the small-scale structure of matter.

Another law of behaviour, very basic to quantum mechanics and related to the fact that electrons are indistinguishable from each other, is embodied in a rule called Pauli's exclusion principle. It states that no two (or more) electrons can occupy the same physical state. This rule determines the way additional electrons are added, to form neutral atoms, as the nuclear charge increases. It leads to the well-known periodicity in chemical properties of the elements and also, in effect, prevents atoms or molecules being squeezed into each other—an interpenetration which would have seemed easy since they are apparently regions almost devoid of matter.

This kind of approach has led to a complete understanding of the behaviour of atoms. Similarly the study of the electron motion in the presence of two nuclei, as in Fig. 2.2(b) can lead, at least in principle, to a description of molecules, so that the basic facts of chemistry can be accounted for. We can calculate in detail some of the properties of chemical compounds, and explain qualitatively many more; though for the large molecules of organic chemistry a full description would be too complicated, and in that area the experience of the practical chemist is still superior to the computer.

Solid bodies are combinations of a great number of atoms or molecules, and here, too, quantum mechanics has given us a very extensive picture of the phenomena. This is not only a question of explaining what we have already observed, but the theory can also be used to make new predictions and with their use design new devices; the transistor and the laser are technologically important examples of this kind.

Even where the general laws are well known, their application to particular situations may take a long time, because of the complexity of the problem. An outstanding example is the phenomenon of superconductivity, by which certain metals suddenly lose their electric resistivity completely when cooled to some very low temperature, usually in the liquid-helium range, so that such a metal becomes, below its critical temperature, a perfect conductor of electricity. This phenomenon was discovered many years before the completion of quantum mechanics in the late nineteen-twenties. Yet the explanation was not given until about thirty years later. By now such superconductors are used commercially for generating strong magnetic fields without using much power, and they may one day be used for transmitting electric power over large distances without loss, though this involves keeping the cables at very low temperature and we have not yet learned how to do so economically.

What I have talked about so far amounts to a complete basis for understanding the structure of atoms, composed of nuclei and electrons, of the electric forces which hold atoms together, and of the behaviour of atoms, and systems of atoms, under the influence of these forces. This is enough for almost all practical applications. However, of the nucleus in the centre of the atom we have so far noted only its mass, which comprises practically all the mass of the atom, and its electric charge. That this is enough for most purposes is because the nucleus is so much smaller than the atom. How much smaller can be seen from Fig. 2.3. Fig. 2.3(a) is a sketch of a heavy atom, say of gold. Its electrons are grouped in spherical shells (more or less like onion shells) which are indicated approximately by the circles in the figure. In the centre is a small dot, and this represents the innermost or 'K'-shell. Fig. 2.3(b) shows the K shell enlarged, and the small dot now visible in its centre is the nucleus.

This small nucleus determines, by its electric charge, the nature of the atom. The fact that gold behaves differently from lead, and both very differently from, say, helium, is due entirely to the different charges on their nuclei, and therefore the different numbers of electrons in a neutral atom.

The lightest atom, containing only one electron, is that of hydrogen. Its nucleus, the proton, is a constituent of all nuclei. What else do they contain? The other nuclear building block was discovered only in 1932. This is the *neutron*, an electrically neutral particle, of approximately the same mass as the proton. All nuclei are made up of neutrons and protons. The charge on the nucleus gives us directly the number of protons in it, and its weight tells us the total number of neutrons and protons.

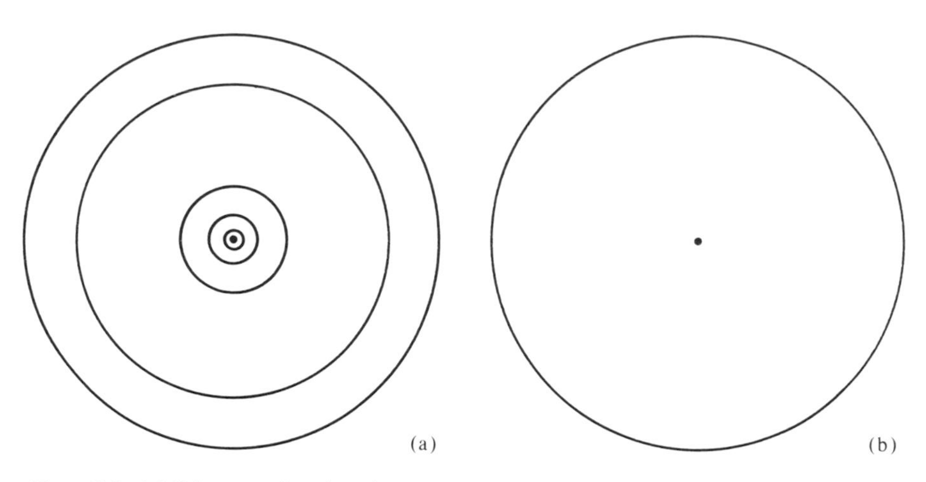

FIG. 2.3. (a) Diagram showing the sequence of electron shells in a heavy atom such as gold. It is roughly to scale and the central dot shows the size of the innermost shell (the K-shell). (b) The K-shell and at the centre, approximately to scale, the gold nucleus.

The structure of nuclei

The structure of the nucleus is very different from that of an atom. Whereas we have seen that the atom, like the Solar System, is very empty, the nucleus is reasonably well filled. Instead of thinking of the electrons occupying a fairly empty space we must visualize the nucleus rather like Fig. 2.4, which shows schematically a helium nucleus, consisting of two neutrons and two protons, their sizes in relation to their distances from each other being as in the figure. This figure of course, leaves out the refinement of quantum mechanics, in which all these particles are associated with waves, because, for more than one moving particle the uncertainty in their position cannot be represented by a simple drawing.

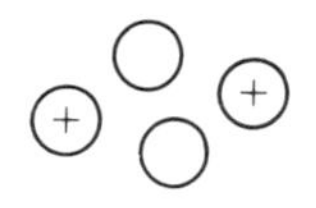

FIG. 2.4. The helium nucleus, approximately to scale, containing two protons and two neutrons.

Next comes the question of the forces which hold the nucleus together. In the case of the atom the attractive force between the nucleus and the electrons is electromagnetic, practically just the familiar electric attraction between opposite charges. So for the atom we do not have to introduce any new kind of force; the familiar law of Coulomb explains the facts. But for the nucleus electric forces will not do, for two reasons. Firstly, of the constituents of the nucleus the protons carry a positive electric charge, while the neutrons carry none. Since like charges repel each other, there is no way of finding a resultant attraction which would hold the nucleus together.

Secondly, the electric forces are much weaker than the forces actually found inside the nucleus. We know the electric forces vary as the inverse square of the distance, like Newton's law of gravitational force, and that means that the energies involved vary inversely as the distance. If electric forces were at work, we would expect nuclear binding energies to be some 100 000 times greater than typical energies in atoms, because distances in the nucleus are that much smaller than typical distances in the atom. But the energies required to remove a particle from a nucleus are much greater even than that.

So there must be a new kind of force at work, the nuclear force, much stronger than the electromagnetic interaction, and it is often referred to just as the *strong interaction* for that reason. Particles passing a nucleus without actually colliding with it are not affected by this new force, but experience only the electric force due to the charge on the nucleus, and this shows that the strong nuclear force is of very short range; at distances greater than a typical nuclear dimension it falls off much more rapidly than the inverse square law. It is no wonder, therefore, that no evidence of such forces has ever been seen in any experiments not involving the structure of the nucleus.

Particles

We have so far mentioned a number of particles and a number of types of force acting on them, and it may be helpful at this stage to summarize what we have learned, and this is done in Table 2.1.

Table 2.1

Name	*Charge*	*Mass (Proton mass as unit)*	*Stable*	*Strong*	*Electromagnetic*
Electron	−1	1/1800	Yes	No	Yes
Proton	+1	1	Yes(?)	Yes	Yes
Neutron	0	1	No (free neutron $\tau \sim 15$ min.)	Yes	Yes
Photon	0	0	Yes	No	Yes

The first particle listed here is the electron. Its electric charge is negative and is given as −1, because it is convenient to use its magnitude as the unit of charge. The unit of mass is taken as the mass of the proton (or that of the hydrogen atom, which, for our purpose, is near enough the same) and on that scale the electron mass is very small, 1/1800 as shown in the table. The electron is stable, it is not affected by the strong interaction, but it does, of course, experience electromagnetic forces.

The proton has a positive charge equal in amount to that of the electron; its mass is one unit by our definition, it is stable, and it is affected by both strong and electromagnetic forces. The neutron has no charge, its mass is approximately that of the proton, it is affected by the strong force and the electromagnetic force (but by the latter only indirectly since although neutral it has an internal charge structure); in the free state, outside the nucleus, it is not stable but decays radioactively in about 15 minutes, a phenomenon to which we shall return later.

The table also includes the *photon*, or light quantum, which, according to the quantum theory, makes up any beam of light or of electromagnetic radiation. It carries no charge, has no mass, is stable when by itself, does not experience strong interactions, but is involved in (in fact constitutes) the electromagnetic interaction.

The table contains the particles most directly evident in the constitution of matter, but we have learned from the theoretical work of Dirac, later verified experimentally, that for each of these particles there exists a counterpart, a so-called antiparticle which is in some way its mirror image. For example, the antiparticle of the electron is the *positron*, a particle just like it but with a positive rather than negative electric charge.

By itself the positron is also perfectly stable, because it carries electric charge, and charge can never be created or destroyed. But it is possible for a positron and an electron to annihilate, because between them they carry no charge, so that their annihilation does not conflict with the law that charge must be conserved; the product of their annihilation is energy, usually carried away by two photons, or γ-rays. Similarly, we can create a positron–electron pair if we can put in the necessary energy. According to Einstein's theory, called special relativity, mass and energy are equivalent ($E = mc^2$). We can work out from the electron mass the energy contained in such a pair and it comes to about 1 MeV. If electrons accelerated by voltages of more than a million volts, or γ-rays of comparable energy, pass through matter, they produce electron–positron pairs. Conversely if a positron meets an electron the two usually annihilate and this is a reason why positrons are not normally seen in nature; matter contains so many electrons that any positron would very soon meet a partner and be annihilated—an extreme form of majority rule!

Similarly there exist antiparticles to every particle shown in our table, and indeed to any particle we shall mention later, except for the photon, which is its own antiparticle.

Particle interactions

There is a simple way, invented by Feynman, of illustrating graphically the interplay between various particles and forces, and an example is shown in Fig. 2.5(a). Here it is understood that time goes vertically upwards and the vertical, or nearly vertical, lines indicate particles: for example the single

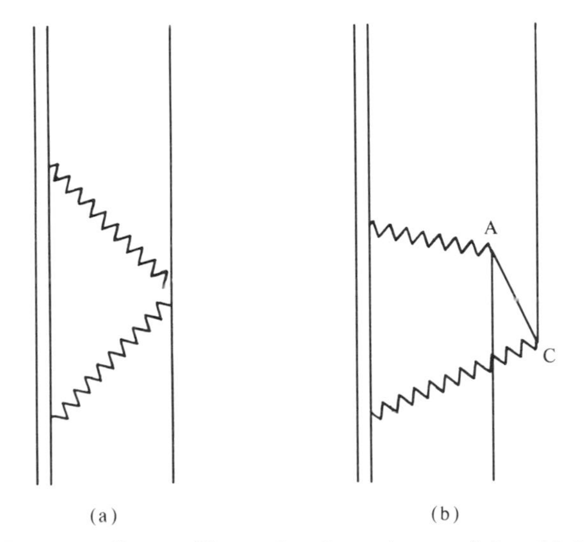

FIG. 2.5. (a) Feynman diagram illustrating the exchange of virtual field quanta (e.g. photons for the electromagnetic field) between two interacting particles (e.g. proton and electron). (b) Exchange diagram including pair creation at C.

lines on the right may indicate an electron and the double line on the left a proton or other nucleus. The wavy lines connecting them are photons, and the electromagnetic interaction, including the electric attraction responsible for binding together a proton and electron to make an atom of hydrogen, can be visualized as an exchange of photons between the two.

A refinement is shown in Fig. 2.5(b) where the electron line doubles back in some part. On the face of it this would suggest going backwards in time, which does not make much sense, but there is a more practical interpretation of the diagram. Following the time development, that is reading from the bottom up, we have initially a single electron line, but at point C two more lines appear, which are interpreted as an electron and a positron, so that at C pair creation has taken place under the influence of the photon line joining at this point. At A the positron merges with the first electron and the two are annihilated, again involving a photon line. The great practical advantage to the physicist of using these types of graphs is that it is much simpler to evaluate the effect of processes of the types in Figs 2.5(a) and 2.5(b) together, than to compute each separately.

We can use this kind of diagram to discuss another important point. If our graph is to represent, for example, a hydrogen atom in its normal state, it might seem unreasonable to talk of photons being emitted or even a pair being created, because we have noted before that, according to quantum mechanics, the electron in the hydrogen atom cannot lose energy, unlike a satellite orbiting around the earth, which can lose energy and crash. If no energy is available, how can photons be produced? The answer is provided by the so-called *uncertainty principle* of quantum mechanics. This has the following consequence for the present situation: although energy is strictly conserved, so that it cannot be created or destroyed, but only transferred from one part of a system to another, you cannot verify this conservation in a short time.

The uncertainty principle, due to Heisenberg, is a fundamental feature of quantum mechanics. It is due to the fact that any observation on an electron, or on any other physical system, necessarily involves some interference with it, for example scattering some light from it so it can be seen. This is true also in classical physics, but there one could in principle make the disturbance caused by the measurement quite negligible. In quantum mechanics, because light consists of quanta, one cannot 'see' a particle by scattering less than one quantum from it, and we cannot reduce the impact caused by it below a certain level. Interactions sharply limited in time must involve photons, or other means of interaction, of high frequency, which carry, and can transmit, greater amounts of energy. Thus a measurement taking no longer than Δt can determine the energy only with an error of ΔE, where $\Delta E\ \Delta t$ is at least of the order of the Planck constant h divided by 2π, which is written $\hbar$. Although we are sure never to find a verifiable violation of the law of conservation of energy, it is yet possible for a system to 'borrow' energy for a short time provided the amount is small enough for the lack of it not to be detectable in the time. The picture is rather like that of a dishonest cashier who borrows money for his own use but returns it before the discrepancy is found out. In the case of the interactions represented by the graphs of Fig. 2.5(a) and 2.5(b) we may therefore think of photons being present for a short time, although there is no spare energy. Such transient particles are called 'virtual' and we can say that these processes involve virtual photons.

The list in Table 2.1 covers many of the particles and interactions which are essential for an understanding of the structure of matter and its properties, yet it is far from complete. One phenomenon which has been known for a long time, and which requires an extension of the list, is the *β-radiation.* The name was given to it by Rutherford who found in his early studies of radioactivity that radioactive substances emitted three types of radiation, distinguished by their deflection in a magnetic field which indicates their electric charge. A sketch of this is shown in Fig. 2.6, where we see three groups of rays in the magnetic field. One, labelled α, is deflected in the same sense as would be a positively charged particle. The rays were later identified as beams of helium nuclei, which are therefore also called α-particles. The undeflected beam, called γ-ray, consists of electromagnetic radiation, high-energy photons. The negative beam, or β-ray, consists of electrons.

More detailed measurements showed that in some elements the α-rays were all deflected by the same amount, indicating they all had the same speed, and so energy. In other cases there were several groups of particles, each group with a sharply defined energy. On the other hand, the β-rays

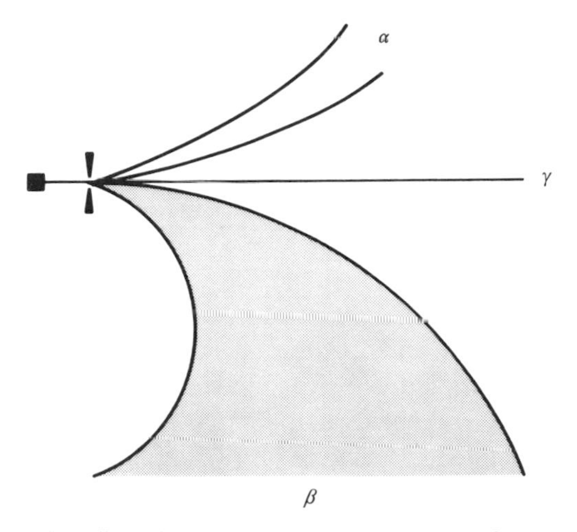

FIG. 2.6. Sketch of radioactive emanations named α-, β-, and γ-radiation by Rutherford.

covered a wide range of deflections continuously, as the figure attempts to indicate, so that they must have a wide range of different energies. This mystified physicists for a long time. We knew that the nucleus must, before the β-emission, have a definite energy, and equally in the final state, after the emission of the β-ray, its energy must be the same for all such nuclei; or possibly it might end up in one of a few states of different, but nevertheless sharply defined, energies. Thercfore the emerging electrons must have all the same energy, or one of a few sharply defined energies, if energy is conserved. The continuous spread of energies suggested therefore that the electron did not carry away exactly the energy the nucleus had lost, and energy was not conserved.

The possibility that we were here witnessing the breakdown of the law of conservation of energy was considered very seriously. We must always be prepared that laws which had been found correct whenever tested, and therefore were believed to be general laws, might be found inapplicable in a new domain of physics not previously explored. However, Pauli pointed out an alternative, namely that in the decay two particles were emitted, the electron and another light particle which we failed to detect. If the available energy was shared between two particles, the electron would take away sometimes a greater, sometimes a smaller share, even though the total energy of both was always the same. This explanation was found to be correct, and the new, very elusive, particle was called the *neutrino*, the Italian diminutive of neutron, since it had to be electrically neutral, but much lighter than the neutron.

This particle is very elusive indeed. It can go through the whole of the Earth, from here to Australia, with a negligible chance of hitting anything on the way. This is true at least at the energies with which neutrinos are emitted from radioactive nuclei. At first one therefore had to be content with circumstantial evidence for the existence of this particle, but the great skill of experimenters, and the existence of nuclear reactors in which enormous numbers of neutrinos are produced, eventually led to the direct detection of these particles. It turns out that at higher energies the interaction of neutrinos with matter increases strongly, and neutrinos produced at the high energies of modern particle accelerators become relatively easy to observe.

The force which gives rise to the production of electrons and neutrinos is very weak. This can be seen from the fact that the average life of a free neutron, which breaks up into a proton, an electron and a neutrino, with an appreciable amount of energy, is about fifteen minutes. This is a fantastically long time on the nuclear scale, and very long compared to other nuclear reactions at comparable energy which are caused by the strong interaction. We witness here a new and extremely feeble force at work, which is called the *weak interaction.* But I should explain, before going on, why it is that for most atomic nuclei the neutrons bound within them are stable and do not decay after only 15 minutes. The reason is to be found in Pauli's exclusion principle, which also applies to protons and neutrons in the nucleus and very strongly inhibits, even prevents, neutron decay because in most nuclei there is no vacant state available for the low-energy proton left behind.

We have thus enumerated the basic phenomena occurring in nuclei, but we need to consider their nature in some more detail, in particular the strong force which binds the neutrons and protons together. As explained before, we do not like action at a distance, and prefer to think that any force of interaction between particles is transmitted by some intermediary, which we call a field. Thus also the strong nuclear force should be transmitted by some field over the distances, however short, over which it acts. Now in quantum theory, with every field there goes a quantum, or particle, like the photon. One could therefore speculate what kind of particle would transmit the strong nuclear force, just as in the situation illustrated in Fig. 2.5 we could think of the electromagnetic interaction as caused by the emission of a virtual photon by one particle and its absorption by the other. So we can use Fig. 2.5 also to illustrate the interaction between a neutron and a proton, or between two protons or two neutrons in the nucleus, with some particle being exchanged between them.

We know that the force is of very short range, and this will follow if the particle being exchanged is not massless, like the photon, but has an appreciable mass. Then its creation takes a good deal of energy, and, as we have already discussed, this energy can be 'borrowed' only for a short time. In such a short time the particle cannot get very far from its source and hence the effect it causes is restricted to a small neighbourhood.

We can estimate the range by the following argument. If the amount of energy that has to be borrowed to provide the intermediate particle is E, it must be 'repaid' after a time $\hbar/E$, as we noted earlier. Since no action can be propagated faster than with the speed of light, c, the greatest distance over which it can have an effect is $\hbar c/E$. To create a 'virtual' particle of mass m, the necessary energy is, by Einstein's relation, $E = mc^2$, so the distance is $\hbar/mc$. The photon has no rest mass, and 'soft', that is long-wavelength, photons can create an influence over large distances, and hence the electromagnetic field can act over an unlimited range. The nuclear force is effective only over a few fermi (the unit of distance equal to 10^{-13} cm, convenient for nuclear sizes) and by the above relation this would correspond to the force being carried by a virtual particle of about one-seventh of the proton mass. This was suggested by Yukawa, in 1935, and a particle of about this mass does exist; it is the *pion*, or π-meson. Energy conservation imprisons these particles unless energy greater than their mass, mc^2, is supplied from outside; this is possible in the collision of two nuclei accelerated to high energy and is a frequent occurrence in the atmosphere which is constantly bombarded by the cosmic radiation, very high-energy particles, mainly protons, arriving from outer space.

We can now revise our Table 2.1 replacing it by Table 2.2. Here the neutrino has been added as well as the pion, of which there are positively and negatively charged, as well as neutral varieties. Amongst the interactions we have added the weak interaction which is responsible for β-decay.

Table 2.2

Name	*Charge*	*Mass (Proton mass as unit)*	*Stable*	*Strong*	*Electromagnetic*	*Weak*
Electron	−1	1/1800	Yes	No	Yes	Yes
Proton	+1	1	Yes(?)	Yes	Yes	Yes
Neutron	0	1	No ($\tau \sim 15$ min.)	Yes	Yes	Yes
Photon	0	0	Yes	No	Yes	No
Neutrino	0	0(?)	Yes	No	No	Yes
Charged pion	±1	1/7	No ($\tau \sim 10^{-8}$ s)	Yes	Yes	Yes
Neutral pion	0	1/7	No ($\tau \sim 10^{-16}$ s)	Yes	Yes	Yes

So by now we know what all matter, including the nucleus, is made of, and we know something about the forces which hold the nucleus together. We are not yet ready at this point to give a full quantitative description of all this, because some details are as yet uncertain, and because the whole situation is rather too complicated for a simple treatment. But, for all we knew when this state of affairs was reached in the late 1940s, this might have been the end of the story but for the details. Nothing further seemed to be needed to make a complete picture.

More particles

However, the discoveries in physics did not stop there. I have in fact not followed the development chronologically because already, before the time I have been talking about, a new particle had been discovered which did not seem to be required for our purpose of understanding the structure of matter, and whose function in the scheme of things is still obscure. This 'unnecessary' particle was the *muon*. It was discovered in the cosmic radiation before the pion, and was first mistaken for the particle predicted by Yukawa to account for the nuclear force. It took some time before C. F. Powell discovered the pion, and the distinction between these two particles was clarified.

That was only a beginning. More new particles were found, produced by the cosmic radiation; a particle accelerator provided by nature, which at first was superior to all man-made accelerators, though by now the artificial ones win for most purposes. One by one further types of particles were identified, and the list started to grow. I shall not attempt to review the history of these discoveries, but I will try to give an impression of the bewildering variety of particles known today. In presenting even some part of the catalogue we need to know the system by which particles are grouped in classes according to their nature and behaviour. One useful distinction is between *hadrons* and *leptons*. Hadrons are particles which can be subject to strong interactions, whereas leptons are those acted upon only by electromagnetic and weak forces. Amongst the hadrons there are two classes: *baryons* and *mesons*.

Baryons are particles which can change into protons, or be made from protons. The background to this is the following. Protons, the nuclei of hydrogen atoms, seem indestructible. We could imagine that the proton and electron of a hydrogen atom might annihilate each other. They carry opposite and equal electric charges, so charge conservation would not be violated, and there is no other obvious conservation law of physics which would prohibit this process. Yet we know that it does not happen in practice. If it could happen even at a very slow rate, so that an atom would live for a time only a little less than the age of the Universe, most atoms would by now have gone up in smoke, the 'smoke' being radiation, and we would not be here. So, if the prohibition on this reaction is not absolute, it must at least be extremely strong. Protons can change into neutrons and vice versa, as we know from the phenomenon of β-decay, with the creation or absorption of leptons, so a neutron is also a baryon. Although we know no basis for it, we introduce a conservation law of 'baryon number' to represent (not explain) this apparent stability of the proton; it is $+1$ for all baryons, -1 for anti-baryons (proton and anti-proton can annihilate), and zero for mesons and leptons.

Mesons, on the other hand, can be created or destroyed singly. An example already mentioned is the pion, which can be produced in the collision of protons or other nuclei at sufficiently high energy.

Yet another distinction arose from the discovery that some of the new particles were rather long-lived, that is long on the time-scale appropriate to such sub-nuclear phenomena, although they were heavy enough to decay into lighter objects without violating the conservation of charge or baryon number. Yet they were easily produced in high-energy collisions so that any prohibition on their decay did not apply to their creation. This paradox was resolved by realizing that there was a new conservation law at work, which allowed two such particles to be produced together. We have seen before that an electron could not be created or destroyed singly, because of conservation of electric charge, but positron–electron pairs are easily produced when the necessary energy is available.

Since the behaviour of these particles appeared strange, the word *strangeness* was introduced for the new conserved quantity, and it followed that one could easily produce only a particle of positive and one of negative strangeness as a pair, together. This law of conservation of strangeness is not absolute, but can be violated by the weak force, so the weak interaction causes the slow decay of strange particles.

With these definitions, let us now briefly look at some lists of particles. Table 2.3 shows some of the baryons which have been found. They come in several categories, and for each the table gives a list of observed masses. In the first category, N, the first entry is 0.939, which is 1 for our purposes, and this is the proton or neutron. They belong to the non-strange particles. The table lists only the lowest known masses and of these only the ones most firmly established.

A second category called Δ also has no strangeness. The first entry there represents the first excited state of the proton, that is, a state to which a proton can be excited by hitting it moderately hard with another particle.

Table 2.3
Baryons

Name	*Strangeness*	*Masses* (GeV/c^2)
N	0	0.939, 1.470, 1.520, 1.535, 1.670, 1.688, 1.700, 1.710, 1.810, 2.190, 2.220, 2.600, 3.039
Δ	0	1.232, 1.650, 1.670, 1.890, 1.910, 1.950, 2.420, 2.850, 3.250
Λ	−1	1.116, 1.405, 1.520, 1.670, 1.690, 1.815, 1.830, 2.100, 2.350, 2.585
Σ	−1	1.193, 1.385, 1.670, 1.750, 1.765, 1.915, 1.940, 2.030, 2.250, 2.455, 2.620
Ξ	−2	1.317, 1.530, 1.820, 2.030
Ω	−3	1.672

We shall not discuss the difference between the N and Δ types, but it has to do with the question of grouping of particles of practically the same mass but different charges. The proton, of charge +1, has only one partner, the neutron, of charge 0. The same pairing applies to all other N-type baryons. To each Δ mass belong states of four different charges, +2, +1, 0, −1.

Then follow the strange baryons, including as the most extreme, the Ω^- with a strangeness of −3. Its decay is very interesting, because even the weak interactions cannot easily cause a change of strangeness of more than one unit at a time, so that the decay of the Ω^- is a three-stage process.

We cannot enter here into all questions of detail, but it will be clear that the picture at this stage is exceedingly complex. Similar complexity is seen in Table 2.4, which lists mesons. Here the practice has been to be more generous with names.

Leptons are shown in Table 2.5. All the ones listed have already been

Table 2.4
Mesons

Non-Strange		*Strange*	
Name	*Mass (GeV/c²)*	*Name*	*Mass (GeV/c²)*
π	0.140	K	0.494
η	0.549, 0.958	K*	0.891, 1.423, 1.780
ϱ	0.770, 1.250, 1.600	Q	1.280, 1.400
ω	0.783, 1.675		
δ	0.970		
S*	0.980		
φ	1.019		
A_1	1.100		
B	1.235		
f	1.279, 1.514		
D	1.285		
A_2	1.310		
A_3	1.660		
g	1.700		

Table 2.5
Leptons

Name	*Charge*	*Mass (Proton mass units)*
Electron	−1	1/1800
Muon	−1	1/9
Electron-neutrino	0	0(?)
Muon-neutrino	0	0(?)
- - -	- - -	- - -
- - -	- - -	- - -

mentioned except for the fact that there are two kinds of neutrino, one associated with the electron, the other with the muon. The broken lines indicate that the table is growing, and there already exists evidence for yet further leptons.

We see, then, that physics had become distressingly complex again. Modern atomic theory reduced the number of basic constituents of matter, from the 92 elements known to the chemist, to the electrons and nuclei, the latter made up of neutrons and protons. We had not completed a full quantitative account of nuclear structure, but a simple basis for understanding the story seemed almost within our grasp.

Instead we were then led to this bewildering new variety of particles not actually part of the structure of matter, but closely related to its constituents and to the forces between them. It is clear that this cannot be the end of the story; there must be some simpler principles beneath this complexity. Where do we have to look for this? It turned out that the scheme of hadrons, which, from our tables, seemed quite haphazard, shows some regularities. The formal structure of this scheme was recognized in 1964 by Gell-Mann and by Ne'eman in terms of a mathematical symmetry scheme called SU(3).

Quarks

The structure of this scheme is exactly what one would find if the baryons and mesons were built of still smaller sub-units. We had given up the idea of dividing everything up into smaller and smaller pieces, but it really does now look as if inside the baryons and inside the mesons we could find new particles, which Gell-Mann called *quarks*, borrowing a term from *Finnegan's Wake*. The original SU(3) scheme required three distinct types of quark but subsequently a fourth was found to be necessary and now it seems there is a fifth and probably a sixth. The first four are shown in Table 2.6.

Table 2.6
Quarks

	Name	*Charge*	*Strangeness*	*Charm*
up	u	⅔	0	0
down	d	−⅓	0	0
strange	s	−⅓	−1	0
charmed	c	⅔	0	1

I shall not discuss the reason why the first two are called 'up' and 'down'. The third is called 'strange', and the last '*charmed*'. Surprisingly, they carry electric charges which are fractions of an electron charge, the 's' quark also carries strangeness and 'c' carries charm, a new quality introduced, like baryon number and strangeness, to account for a new conservation law which prohibits processes that might otherwise take place. Charm is again an arbitrary name for a quantity we cannot otherwise describe but it 'worked like a charm' to remove difficulties in the theory.

Baryons, which include the neutron and the proton, are made of three of these quarks, and this makes their charge always come out to be a whole number of units or zero. Each meson is made of one quark and one antiquark. As we saw earlier, to each particle there exists its mirror image, or antiparticle, and that goes also for the quarks. As each quark has to have a baryon number of 1/3, since three of them can make a baryon, antiquarks must therefore have a baryon number of $-1/3$, and the resulting baryon number of the meson is zero, as it should be.

This scheme leads to many features of the hadrons which are observed, and there is strong evidence, growing all the time, that the quark model gives a good description of the hadrons. Yet nobody has ever seen a quark. Because of their fractional charge they would leave unmistakable tracks if they turned up in any experiment. They have been looked for very carefully, but without success.

The present belief is therefore that quarks cannot exist in freedom, they are confined to the interior of the hadrons. While they are inside they move like ordinary particles, and good progress has been made towards a mechanical description of their behaviour, and thus of the behaviour of the hadrons of which they form part. But, according to this view, it is impossible to separate them.

Forces

Now again the question arises as to the force which holds the quarks together and determines their motion. It is again true that such a force, which must be strong since the known strong interactions arise from it, must be transmitted by some field which will have quanta, or particles, associated with it. These have been given the name *gluons*, because in a sense they form the glue that binds the quarks. The gluons complete, for the present, the list of basic units required for the hadrons and their strong interactions.

Study of the weak interaction suggests the existence of yet another important particle. To see this we recall that the other interactions, the electromagnetic and strong interactions (and even gravitational), can be represented by diagrams of the type of Fig. 2.5 in which at each junction, or vertex, only three lines meet. As a typical weak interaction consider the decay of a neutron, where we have the situation sketched in Fig. 2.7(a). The incoming neutron decays into a proton, and electron, and a neutrino, and if this all happens in one elementary process four lines would have to meet at one point. Such a fourfold vertex presents great difficulties for the theoretical description and the question arises whether the true picture does not perhaps look like Fig. 2.7(b), where the wavy line denotes some new kind of particle, called an *intermediate vector boson*.

If this is heavy, the energy necessary for its creation would not be available, but it could be 'borrowed' and if the mass is high the weak force will extend only a very short distance in space. The process would therefore be hard to distinguish from that pictured in Fig. 2.7(a), in which everything happens at a point.

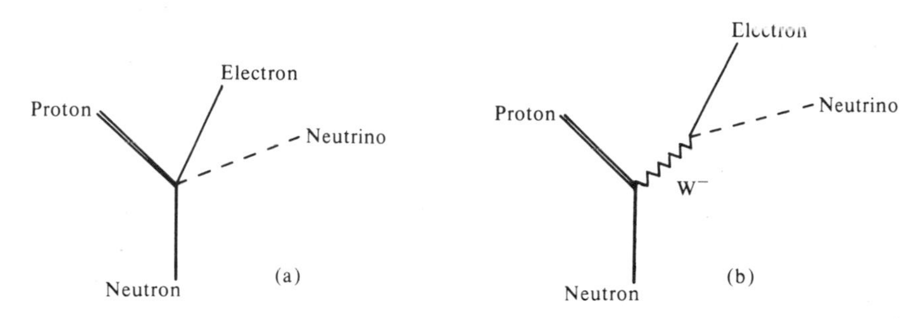

FIG. 2.7. (a) Diagram of neutron decay by a weak interaction in which all four particles connect at a single vertex. (b) Neutron decay involving an intermediate vector boson, W.

The effective strength of this weak interaction, that is, for example, the rate of decay of the neutron, would depend on the strength of the coupling acting at the vertices in the figure. But it also depends on the mass of the new particle, because if it is heavy it is difficult to produce, and the strength of the effect represented by the diagram is reduced. We can therefore make the interaction weak either by reducing the basic interaction at the vertices or by raising the mass of the intermediate vector boson.

This opens up the possibility of a universal theory of weak and electromagnetic interactions; in this theory the strength of the interaction of the vertices in Fig. 2.7(b) is almost the same as for the electromagnetic interaction we were talking about in connection with Fig. 2.5. This then fixes the mass of the intermediate boson at about 100 GeV if the theory is to account correctly for the weak interactions, and while it is understandable that this particle has not yet been seen in experiments, future accelerators should before long reach the energy required to produce it. While there is as yet no final verdict on this universal theory, the chances of it being right are excellent.

I have touched upon a large number of things and hope to have given an impression of the way in which the physicist describes the world in terms of different classes of particles which make up the objects we see around us, and the forces which make them behave as they do.

Returning to the basic types of force, we have, in order of strength, the strong, electromagnetic, weak, and gravitational forces. The electromagnetic forces are somewhat weaker than the strongest, but still appreciable—we use them every day. The weak forces are enormously weaker in their effects but, as we have seen, they may still be basically connected with the electromagnetic interaction. About the gravitational force I have not talked at all, except in the beginning, because it is so weak that in all atomic and nuclear problems its effect is completely negligible. But this gives rise to the question: why is it that this so fantastically weak force has been known the longest, and that we experience it more in our daily lives than any of the others? There is a good reason for this. It is that the gravitational force is cumulative. The pull of gravity exerted on us by the Earth is the combined effect of the attraction due to all the atoms of which the Earth consists. These forces act all in the same direction. There is not positive and negative gravity, as there is positive and negative electricity. Electric forces cancel out unless there is a large net charge, and usually large bodies are electrically neutral, that is they contain as many positive as negative charges. The strong interactions are, as we have seen, of very short range and therefore can never add up in their effect on a given object.

This statement about gravity is true even in its effect on antimatter, which is made of antiparticles. There has been some science fiction based on the idea that antimatter might experience a gravitational repulsion, instead of attraction. However, in reality gravity acts the same way on antimatter. This consistency of sign and the cumulative nature of the force are the reasons why the weakest of all forces caught our attention first.

3

Manifest and Hidden Symmetries

C. H. LLEWELLYN SMITH

Symmetry is a continuous source of fascination and pleasure. We all enjoy patterns, whether in natural objects such as crystals, or man-made ones like Persian rugs or mosaics. Although we all have a clear idea of what is meant by symmetry, there is an excellent operational definition due to Hermann Weyl: an object is symmetrical if there is something you can do to it such that when you have finished it looks the same as it did before. Consider, for example, the snowflakes in Fig. 3.1 which are clearly symmetrical; there is something we can do, namely rotate them through 60°, so that afterwards they look the same as they did before. In addition to this rotational symmetry they also have reflection symmetry about certain lines; for example, after reflection about a vertical line they again look the same as they did before.

Rotations and reflections are spatial operations but objects can also be symmetrical under operations which are not spatial or are partly spatial and partly non-spatial. Consider the picture by Escher shown in Fig. 3.2. It is certainly not rotationally invariant nor is it the same as its mirror image. However, it is symmetrical under the combined operation of a reflection and the interchange of black and white.

We shall be more concerned with what physicists call the symmetries of the laws of the nature than with the symmetries of objects. What do we mean by saying that a law has a symmetry? Weyl's definition is still valid: a law of nature is symmetrical if there is something you can do to it such that when you have finished it looks the same as it did before. Let me immediately give two examples.

1. The laws obeyed by the nuclear force look the same if we swop neutrons and protons. How do we know this? We know it because the properties of nuclei look the same if we swop neutrons and protons. For example, the properties of the lithium-7 nucleus, which contains three protons and four neutrons, are almost the same as those of the beryllium-7 nucleus,

FIG. 3.1. Snowflakes. (Barnaby's Picture Library.)

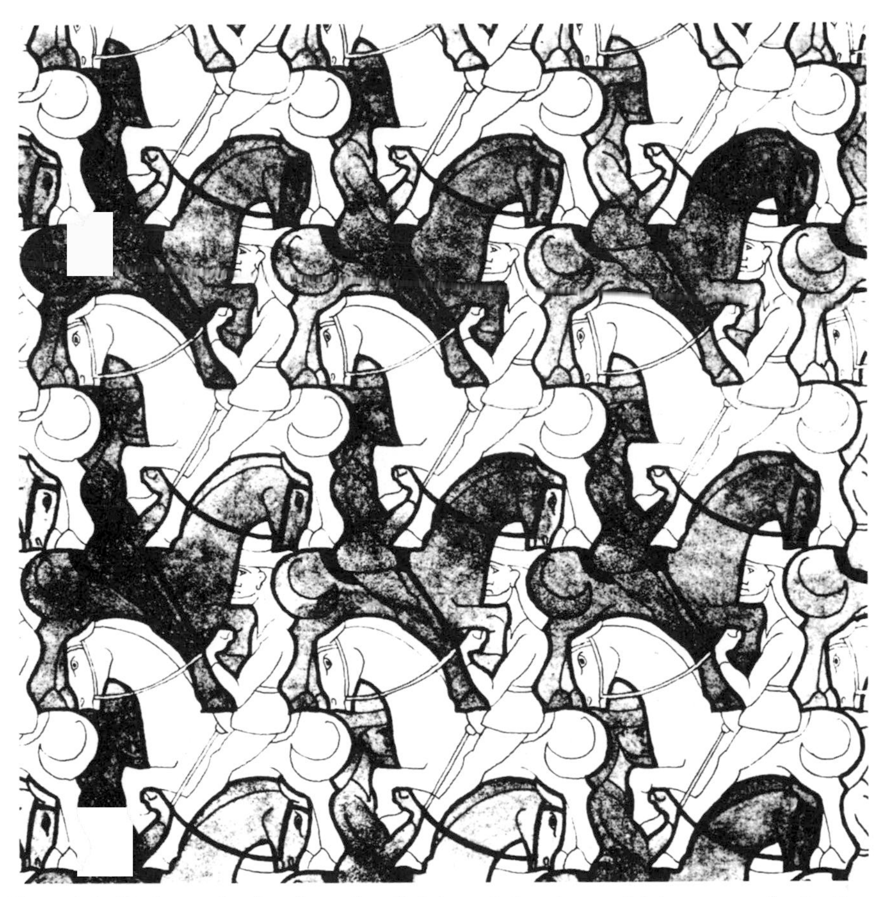

FIG. 3.2. Escher—Study of regular division of the plane with horseman (collection Haags Gemeentemuseum—The Hague).

which contains four protons and three neutrons (we shall see the evidence for this later). They are not exactly the same because the proton, which is electrically charged, and the neutron, which is not, respond quite differently to the electric force and although the influence of electromagnetism on the properties of nuclei is small, it is not negligible. Allowing for the electromagnetic differences, however, the properties of the lithium-7 and beryllium-7 nuclei do seem to be the same. We therefore conclude that the nuclear force is symmetrical under the interchange of neutrons and protons.

2. The laws of physics are the same at different places, or equivalently, the results of experiments are the same if they are carried out at different places, a property known as translational invariance. For example, the force between a planet and a star depends only on the distance between them and not on where they are. If both are moved by an equal amount,

their behaviour will be exactly the same as it would have been in their original position (provided the change has not brought them under some external influence, such as another star). It follows that the underlying laws must be unchanged if we shift everything to which they refer by an equal distance.

Why are we interested in the symmetries of physical laws? The discovery of symmetries is pleasing for two reasons. First, as in the case of objects, where symmetry is synonymous with harmony and is associated with elegance and aesthetic pleasure, so in the case of natural laws, symmetry is generally associated with mathematical elegance. Second, the more the symmetry the greater the predictive power. For example, knowing that nuclear forces are symmetrical under the interchange of neutrons and protons we can predict the properties of one nucleus given those of another. It is clearly much simpler to have just one law for the behaviour of both neutrons and protons, rather than a separate law for each, and to have laws which are valid throughout the whole Universe rather than different laws at different places. In fact symmetry can be a tight strait-jacket. The higher the degree of symmetry, the more tightly the laws are circumscribed; correspondingly, the less we have to put into our laws from experiment in the way of measured constants of nature, and the more we can get out in the way of predictions.

I shall refer to symmetries such as translational invariance and the symmetry of the nuclear force under the exchange of neutrons and protons, which are clearly seen in nature, as manifest symmetries. Manifest symmetries must obviously be built into physical laws. Surprisingly perhaps, the converse is not true. Symmetries which are built into physical laws need not show up clearly in nature. In other words, symmetrical laws can in certain circumstances give rise to asymmetrical phenomena. This is often misleadingly known as spontaneous symmetry breaking. I shall refer to it as hidden symmetry. The idea that nature may have hidden symmetries is exciting because it makes it possible to seek a common (simple, symmetrical) origin for apparently quite unconnected phenomena. For example, the weak, electromagnetic, and strong forces appear to have nothing in common. Nevertheless, it is possible that there is a hidden symmetry between them so that there is just one force law giving rise to weak, electromagnetic, and strong phenomena, making it possible to relate their very different properties.

Manifest Symmetries

Symmetries of space and time

Symmetries associated with space and time are easy to visualize. We believe that the results of experiments are independent of where and when they are performed and of the orientation of the apparatus. Of course this is not literally true. If an experiment is susceptible to the influence of gravity, the results will obviously be different if it is performed on the Moon rather than on the Earth. There are many other effects, such as temperature and the magnetic field of the Earth, which may influence experiments and do depend on time, place, and orientation. However, it is part of the art of experiment to remove these effects or discount them. Having allowed for the obvious effects of the local environment, we believe that the symmetries hold good. This belief in reproducibility is actually taken for granted in the way science is done, although I certainly don't agree with Wigner who asserts that science would be impossible if it were not true. Of course it is possible that the laws of physics are different in far regions of the Universe or that they were different in the early Universe, just after the 'Big Bang'. However, there is no need for this hypothesis, and the evidence suggests that the laws of physics are in fact essentially the same throughout the Universe and have remained unchanged with the passage of time.

Each of these symmetries of space and time has a very interesting consequence. It leads to something called a conservation law. We say that a quantity is conserved if it does not change in time. For example, translational invariance implies the law of conservation of momentum, which states that the total momentum of a system does not change in time. Thus if the two ice skaters discussed by Sir Denys Wilkinson in chapter 1 (Fig. 1.6) collide, the sum of their momenta after the collision will be the same as it was before. Not only does translational invariance imply momentum conservation, but the converse is true. If momentum is conserved then the underlying laws must necessarily be translationally invariant. Similarly rotational invariance, the statement that the results of experiments do not depend on orientation, implies that angular momentum is conserved and, vice-versa, angular momentum conservation implies rotational invariance. Invariance under translations in time implies energy conservation and vice-versa. These are examples of a very general and, as we shall see, very powerful connection between symmetries and conservation laws. Most symmetries imply conservation laws and conservation laws are always associated with an underlying symmetry.

Unfortunately I cannot make this general connection plausible using common sense arguments as it is not necessarily true in the common sense world of classical mechanics, which is a framework for applying the laws of physics to phenomena on scales which are large compared to atoms. But it is true in quantum mechanics, which we believe gives a correct description on all scales, and in those versions of classical mechanics which can be derived from quantum mechanics.

The important role of conservation laws can be illustrated by the example of angular momentum conservation which follows from rotational invariance or, if you prefer, the statement that there is no preferred direction in space. We shall not consider the law of angular momentum conservation in its full generality but limit ourselves to the special case of a planet orbiting the Sun. In this case conservation of angular momentum is actually equivalent to Kepler's second law. Johann Kepler's three empirical laws or rules of planetary motion, which he published in the early seventeenth century, were based on painstaking measurements of the positions of planets made over many years. The second law states that the orbits of planets sweep out equal areas in equal times, as illustrated in Fig. 3.3. Now this property of the motion is equivalent to angular momentum conservation and this we know follows solely from rotational invariance. To understand Kepler's second law we need know nothing about the particular form of the law of gravity, the force which holds the planet in its orbit around the Sun, except that it is rotationally invariant. There are three important lessons to be learned from this:

1. Since Kepler's law is known to be true, the laws of gravity must be rotationally invariant. This is one example of many cases in which conservation laws help us to identify symmetries which then constrain the form of physical laws, in this case the law of gravitational force.
2. A clear understanding of the connection between conservation laws and symmetries is important because it allows us to distinguish whether a prediction depends on a particular form of a law or on general properties of the phenomena to which it relates. Thus although Kepler's law does follow from Newton's inverse square law of gravitation, we learn that this is not a complete test of the inverse square law but only of the fact that it is rotationally invariant.
3. Having deduced that the laws of gravitation are rotationally invariant from this simple case, we can then use the general form of the angular momentum conservation law to help us in more complex situations, for example in the case of a planet orbiting a double star. In fact symmetries

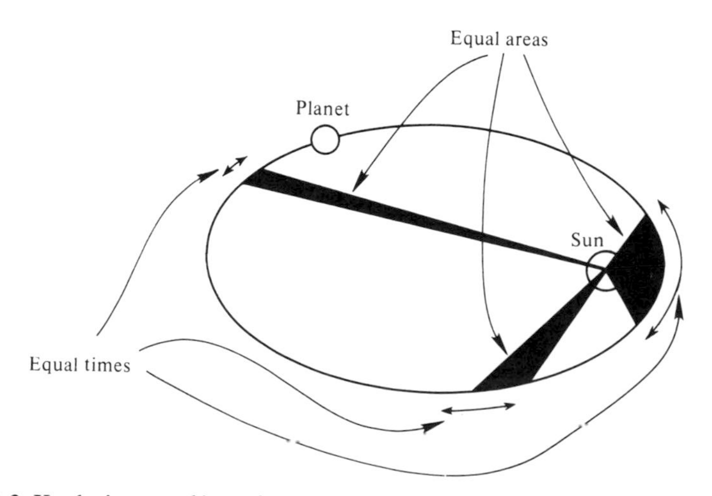

FIG. 3.3. Kepler's second law of planetary motion: equal areas are swept out in equal times.

and conservation laws are always very helpful in applying old laws to new situations.

Up to now we have discussed symmetries which are thought to be exact. Next we consider a symmetry which is either broken or hidden: reflection or mirror symmetry (or parity as it is called by physicists). If mirror symmetry were exact it would be impossible to tell whether a film of an experiment had been made directly or by filming a mirror in which the experiment was reflected. Equivalently, there would be no absolute distinction between right and left. Suppose you were in telephone contact with beings in a distant galaxy. Having established a common language (a soluble problem) you begin to describe yourself. You tell them that you are seventeen thousand million hydrogen atoms high and so on. You then tell them that your heart is on the left. Left? What's that? Unless you are already familiar with what comes next, you will find that you cannot think of a way to make an absolute distinction between left and right. In fact, until 1957 it was thought that there was no distinction. In that year, however, it was learned that although mirror symmetry seems to hold exactly for the strong, electromagnetic, and gravitational forces, it is badly broken in certain weak processes, which had not been carefully studied previously.

The principle of the experiment which demonstrated the violation of mirror symmetry is shown in Fig. 3.4. It consisted of observing electrons

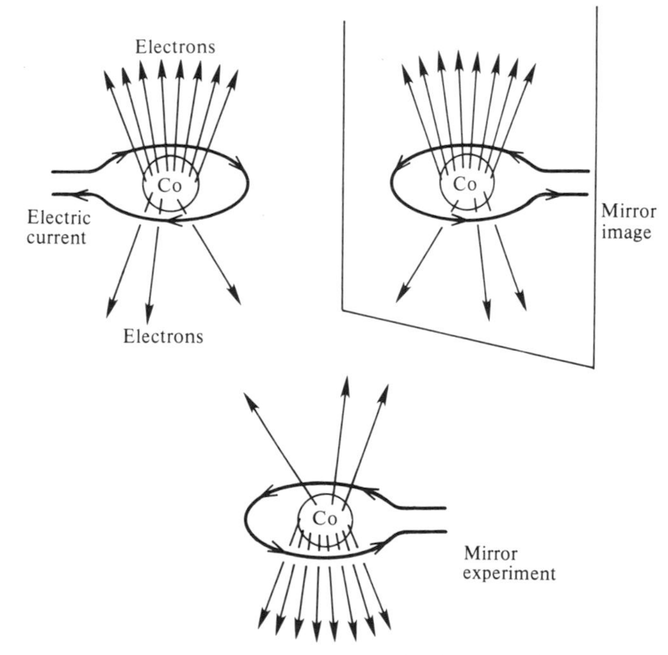

FIG. 3.4. The violation of mirror-symmetry in the β-decay of cobalt. The current in the coil generates a magnetic field which, at a very low temperature, aligns the cobalt nuclei through interaction with their nuclear magnetic moment.

which are emitted in the radioactive decay of the cobalt-60 nucleus. A sample of cobalt, cooled to very low temperature, was surrounded by a coil carrying an electric current in the direction indicated by the arrows. It was found that if the experiment was set up in the way shown in Fig. 3.4, more electrons were emitted upwards than downwards. Actually we should not say upwards and downwards since gravity has nothing to do with it. The correct characterization of the situation is that more electrons were emitted on the side of the coil from which the current is seen to flow clockwise round the loop. Fig. 3.4 shows a mirror image of the experiment and also an actual 'mirror-experiment', that is, one set up so that the apparatus, in particular the direction of the current in the coil, is the same as in the mirror image. In the mirror image, the majority of the electrons are also emitted upwards. In the *mirror-experiment*, however, they would go downwards. This can easily be seen by recalling that the correct characterization of the result of an actual experiment is that more electrons are emitted on the side of the coil from which the current is seen to flow clockwise, and that is downwards in the case of the mirror-experiment. Thus we can clearly tell the difference between the mirror image of the original experiment and an actual mirror-experiment and this constitutes a violation of mirror symmetry. It allows our distant friends to learn what we mean by clockwise, and hence what we mean by left and right, by setting up the cobalt-60 experiment and observing the direction in which the majority of the electrons are emitted.

Actually it is really also necessary to use another experiment to ascertain whether they are made of what we refer to as matter or antimatter, since the cobalt-60 experiment is symmetric under a simultaneous exchange of left and right and of matter with antimatter, just as the Escher drawing in Fig. 3.2 is symmetric under a simultaneous reflection and an interchange of black and white. However, this symmetry, which is known as CP invariance, is violated in certain rare decays of the K meson discovered by Cronin, Fitch, and collaborators in 1964. These decays can be used to make an absolute definition of matter and antimatter. As John Ellis describes in chapter 6, CP violation is crucial to current speculations on the origin of matter in the early Universe.

For a long time it was thought that the violation of mirror symmetry and CP symmetry in weak decays implied that they are not respected by the laws of physics. However it is now recognized that this need not be so; these symmetries could be exact but hidden. Whether or not this is the case is at present unknown.

Many biological phenomena show large violations of mirror symmetry. For example, the double helix of DNA is always right-handed. It is generally thought (although it has never been proved) that once a sufficient majority of specimens of a definite handedness develops in a biological material, it would inevitably increase and the specimens of the other handedness would die out altogether. It is not clear what initial majority would be sufficient. Nor is it known whether an initial majority would be generated predominantly by random fluctuations, which could equally well be left or right handed, or (which seems less likely to me) by the influence of the weak force, in which case the asymmetry of biological material would be determined by the asymmetry of the weak force.

Now let us consider symmetry under changes of scale; that is, we ask whether the results of experiments would change if their dimensions were scaled up by, say, a factor of 2. This question is actually quite old. It was posed and answered by Galileo in his wonderful book, *The Dialogue Concerning Two New Sciences*, in which he showed that a person cannot be scaled up indefinitely. Suppose your dimensions were doubled, so that you were twice as high, twice as wide, and twice as broad. You would be eight times as heavy. However, the cross-sectional area of your bones, and hence their strength, would only be four times as big. If your dimensions were doubled again, you would have 64 times your original weight but your bones would only have eight times their original strength. Sooner or later you would obviously collapse under your own weight. This is a gravitational phenomenon and to discover whether the laws of gravity are symmetrical under changes of scale, we really ought to increase the size of the Earth at the same time. However, increasing the dimensions of the Earth will only make you collapse sooner. The fact that weight increases faster than strength when structures are scaled up is known as the cube–square law by engineers. Galileo goes on to point out that 'if the size of a body be diminished, the strength of that body is not diminished in the same proportion; indeed the smaller the body the greater its relative strength. Thus a small dog could probably carry on his back two or three dogs of the same size; but I believe that a horse could not carry even one of his own size.' This is illustrated in Fig. 3.5, which is taken from the first (1939) edition of *Superman*. However, the picture, and especially the heading 'Scientific explanation of Superman's amazing strength', shows that the authors believed that men ought to be able to do as well as ants and grasshoppers if only they could find the knack. Clearly they had not read Galileo.

Internal symmetries

I have already mentioned the simplest example of an internal symmetry, the symmetry of the nuclear force under the interchange of protons and neutrons. Some of the evidence of this is exhibited in Fig. 3.6, which shows the energies of the excited states of the lithium-7 and beryllium-7 nuclei (these are states in which the protons and neutrons are moving, vibrating and/or rotating, in the various ways allowed by quantum mechanics). Apart from some confusion with the very highly excited states, the spectra are essentially identical. This clearly shows that there is (approximate) symmetry under the interchange of protons and neutrons.

This symmetry was recognized in the 1930s when protons and neutrons

FIG. 3.5. Authors of *Superman* wrongly assumed scale invariance in a gravitational environment.

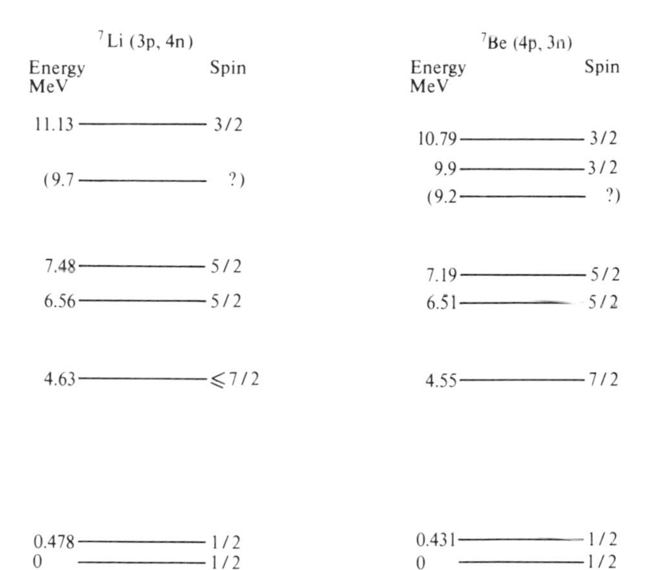

FIG. 3.6. Energy levels of the lithium-7 and beryllium-7 nuclei. Ignoring small effects due to the proton's electric charge, the two sets of levels are the same, demonstrating the symmetry of the strong nuclear force to interchange of proton and neutron.

were the only sub-nuclear particles known (it goes under the name of isospin symmetry). In the 1940s and 50s a host of other particles were discovered and so it was natural to look for larger symmetries connecting these new particles to each other and/or to the proton and neutron. It was immediately clear that such symmetries could only be very approximate; otherwise they would have been obvious, like the symmetry between lithium and beryllium. This made them difficult to find. However, a higher symmetry, with the technical name SU(3), was recognized in 1961. This symmetry groups particles into families with one, eight, or ten members. Its significance is similar to that of the classification of chemical elements in the periodic table, discovered in the nineteenth century. Like the periodic table, SU(3) symmetry suggests the existence of sub-structure. It quickly led to the quark model, according to which all sub-nuclear particles are made of a small variety of more fundamental entities called quarks, bound together in different ways. We shall discuss the quark model first to see how it accounts for the data and implies SU(3) symmetry, but you should remember that the historical process was the reverse.

Until 1974 three quarks were needed. They are distinguished by two attributes called Y and I_3, which have the numerical values shown in Fig. 3.7(a); their counterparts, the antiquarks, are displayed in Fig. 3.7(b). Instead of using the labels Y and I_3 we could have used the electric charge Q, the strangeness S, and the baryon number B (which is equal to 1/3 for quarks

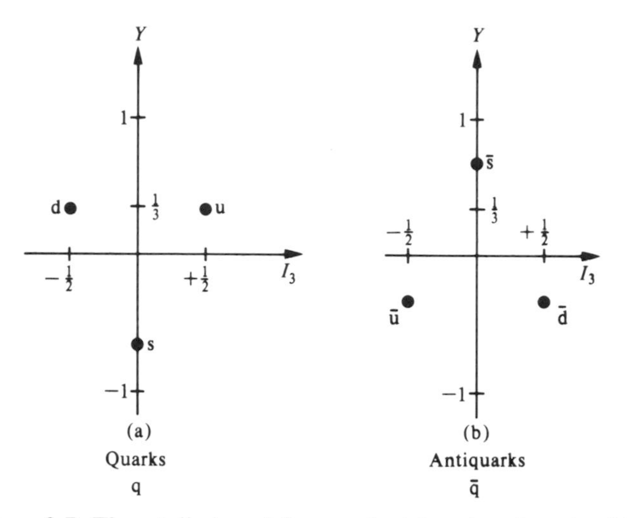

FIG. 3.7. The attributes of the quarks (a) and antiquarks (b).

and $-1/3$ for antiquarks). In fact $Y = S + B$ and $I_3 = Q - Y/2$. There are two ground rules:

1. Although it is possible to create or destroy a quark–antiquark pair (with net attributes zero) it is impossible to make or destroy a single quark or antiquark. Thus the net numbers of quarks minus antiquarks is conserved. This is equivalent to the law of baryon number conservation (we now believe that this law may not be exact, but the fact that the lifetime of the proton is at least 10^{30} years shows that the failure must be quite negligible for almost all practical purposes).
2. Quarks retain their identity, except that very occasionally, only under the influence of the weak force, an s-quark can be transmuted into lighter particles (s → u + leptons or s → u + dū). Therefore, as far as the strong and electromagnetic forces are concerned, the net number of strange quarks minus anti-strange quarks is conserved; this is equivalent to the law of strangeness conservation which is observed by these forces.

What has this to do with symmetry? It seems that u, d, and s quarks respond to the strong force in exactly the same way. However, this symmetry is broken by the fact that the s-quark is substantially heavier than the u and d quarks, which have almost the same mass. Having discussed the ingredients, we can now attempt to make sub-nuclear particles out of quarks and antiquarks.

First, consider mesons, particles which can be created and destroyed singly. They are supposed to be made of a quark and an antiquark. Each can be chosen in three ways, giving the nine possible combinations shown in Fig. 3.8(a) (physically they behave as one family although mathematically they form an SU(3) family of 8 plus a family of one). Imagine the quark and antiquark to be two balls tied together with a spring. The system can vibrate

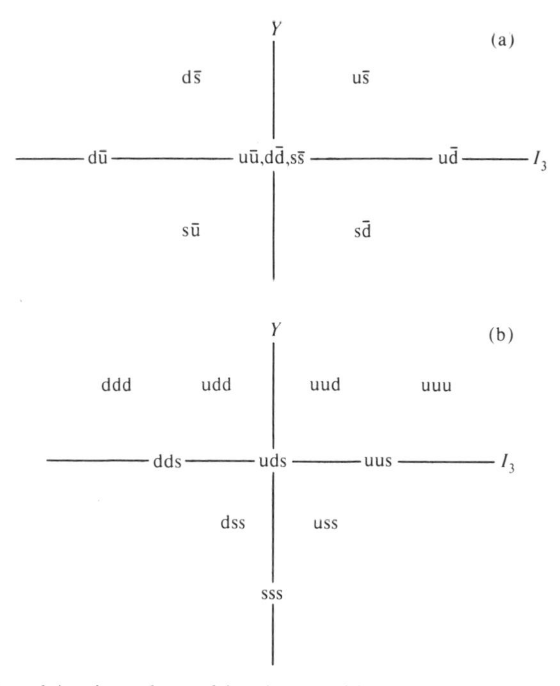

FIG. 3.8. (a) Quark/antiquark combinations making up a set of nine mesons. (At the centre of the diagram there are three meson states formed by taking different linear combinations of the pairs indicated.) (b) The three-quark combinations forming baryons.

and rotate but only with certain special frequencies allowed by the rules of quantum mechanics; in addition, the quark and antiquark can each spin like a top in one of two directions. The energy of the system, and hence its mass according to Einstein's relation $E = mc^2$, will depend on the state of vibration and rotation and orientation of spins. For each possible state of internal motion we would therefore expect to find a family of nine mesons with (roughly) the same masses and other properties. This is exactly what is observed. The lightest family of mesons, the so-called spin zero mesons, which have no internal motion and spins that cancel is shown in Fig. 3.9(a). There is not an exact symmetry between these particles, because the s-quark is heavier, and hence the K-mesons, for example, are heavier than the π-mesons.

Next consider baryons, particles which cannot be created or destroyed singly, which are supposed to be made of three quarks. As for the mesons, to describe a baryon we must now specify the state of motion and spin for the first, second, and third quark. The simplest possibility is that the three states of motion and spin are the same. In this case the order in which quarks of a given type are assigned to these states does not matter so that, for example, the combinations uud, udu, and duu are the same. The ten possible combinations of types of quark shown in Fig. 3.8(b) will then make a family of

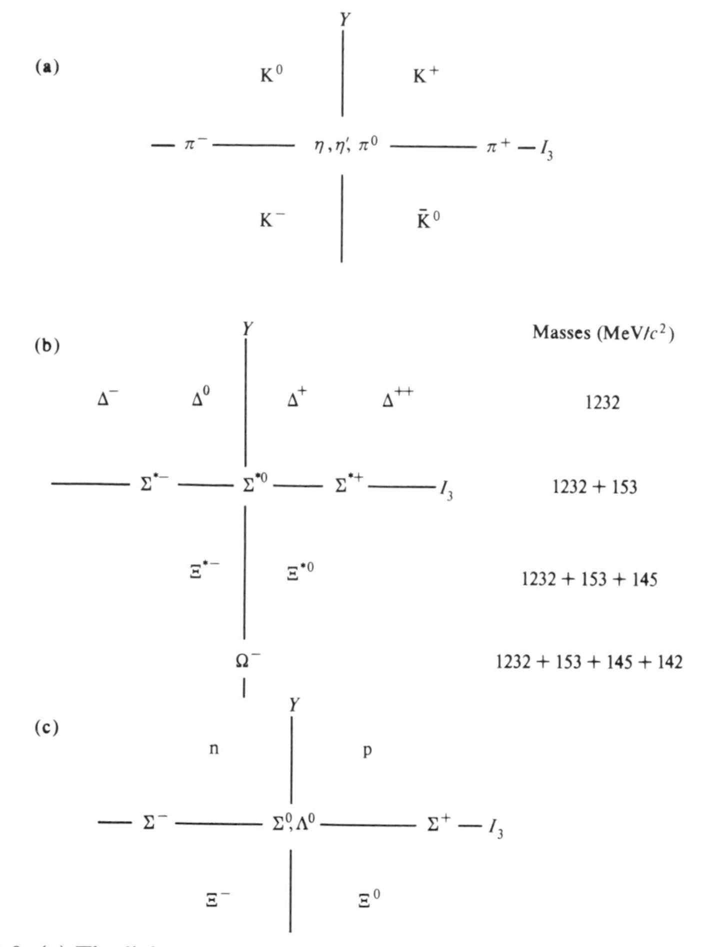

FIG. 3.9. (a) The lightest spin 0 mesons. (b) The lightest spin 3/2 baryons. (c) The lightest spin 1/2 baryons.

ten particles. Such a family, shown in Fig. 3.9(b), is observed. This is the lightest family of ten in which there is no internal motion. As in the case of meson families, the symmetry between these ten particles is not exact but it is broken in a systematic way. The masses shown in Fig. 3.9(b) are directly explained by the hypothesis that the s quark is heavier than the other two.

Now the Pauli exclusion principle, which prevents two electrons being in the same state of motion and spin, also prevents two quarks of the same type, for example two u's, being in the same state. But it seems that the way we built a family of ten baryons violates this principle. This difficulty was resolved by introducing the hypothesis that quarks have an additional hidden attribute, called 'colour', which allows us to choose two u quarks of different colour and put them into the same state of motion and spin. There is now other convincing evidence for the existence of colour which will be discussed by Donald Perkins in chapter 4.

When the states of motion and spin of the three quarks are not all the same the situation is more complex. It turns out that in this case the quarks form families with eight members or just one member. The lightest family of eight, which contains the proton and neutron, is shown in Fig. 3.9(c).

So the quark model accounts for the observed patterns of particles. Historically, of course, the patterns were found first (or rather just parts of the patterns). This suggested an approximate symmetry which later led to the notion of quarks. Quarks not only explain the pattern of particles but many other properties and there is no doubt that the quark model is true.

There is just one cloud on the horizon. Although only three quarks were needed before 1974, at least five quarks are now known. The fourth and fifth quark are much heavier, and therefore particles containing them are much harder to produce and to study and were only discovered relatively recently. It is quite possible that there are many even heavier quarks awaiting discovery. The old simplicity is lost. The reason for the proliferation of quark species is quite unclear. Could it be that quarks themselves have substructure? There is no hint of this in their properties. This proliferation of quark species is one of the greatest puzzles confronting us today.

Hidden Symmetry

The earliest discussion of hidden symmetry is often attributed to Buridan, the fourteenth century French philosopher. His celebrated ass is shown in Fig. 3.10. There are many things to be learned from Buridan's ass, one being that

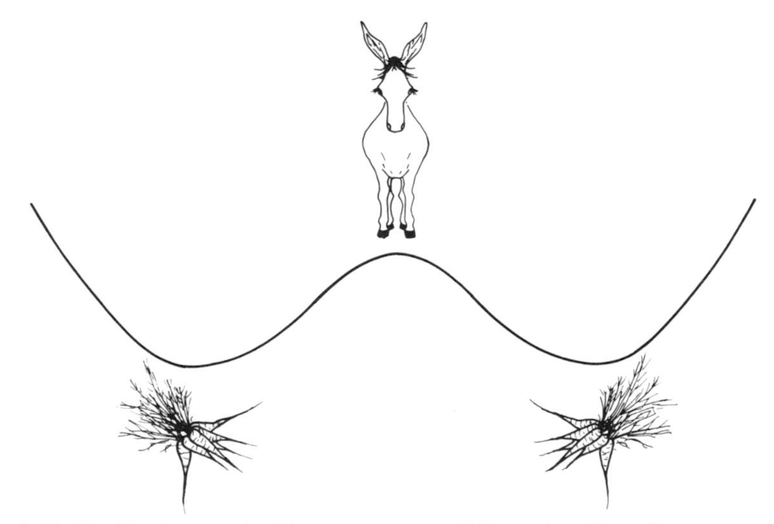

FIG. 3.10. Buridan's ass poised between two equidistant bunches of carrots, starving in a perfectly symmetrical world.

misattribution of ideas is not a new phenomenon; the ass is apparently nowhere to be found in Buridan's writing and it is doubtful if he invented it. In any case, the ass stands poised equidistant from two bunches of carrots, a situation with perfect symmetry about a line drawn through the middle of the picture. The ass is supposed to have died unable to make up its mind which way to move. In fact, of course, it will go to one bunch of carrots or the other, thereby breaking or hiding the symmetry of the situation. I drew the ass on a hill and the carrots in a valley so that I could easily change to a slightly more relevant example. Replacing the ass by a ball (Fig. 3.11), we again have a situation of perfect symmetry. However, it is unstable: the ball will roll into one valley or the other. There is still symmetry in the sense that the ball could equally well roll into either. However, in fact it will roll into just one and the symmetry of the initial situation will then be hidden.

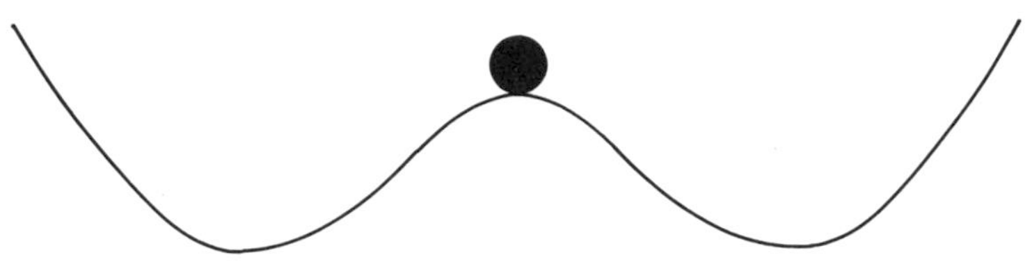

FIG. 3.11. A perfectly symmetrical world, but the situation is unstable; the ball will roll into one or other of the valleys and the symmetry will then be 'hidden'.

The idea of hidden symmetry in particle physics is that the whole Universe is, in an extremely loose sense, analogous to this example. The underlying laws are symmetrical but the symmetrical solutions are unstable. The whole Universe therefore goes into an asymmetrical state, like the ball in the valley. If this is true we live in an asymmetrical environment and cannot directly see the symmetries of the underlying laws. What is the asymmetrical environment in which we live? It is the vacuum, the state of lowest energy in which nothing seems to happen. You may think that the vacuum is empty but we will see that it is in fact seething with activity and may become asymmetrical.

A magnet provides a slightly more realistic example. Magnetism is an atomic phenomenon. Some atoms behave like tiny magnets, and in certain circumstances the forces tend to align each of these atomic magnets with its neighbour. In this case, they will all line up and the whole system will become magnetic. Consider two atomic magnets, indicated by arrows in Fig. 3.12. When they are anti-parallel the situation is unstable and one of them will turn over. There is rotational invariance and we cannot predict which will turn over or in what direction they will line up. However, line up in some direction they will and the rotational symmetry will be hidden. A system of many atoms (Fig. 3.13) also has rotational invariance; no special direction is singled out by the laws which govern it and we cannot predict the direction in which the magnets will point. But they will become aligned in some (random) direction and the symmetry will be hidden. If you were a magnetic

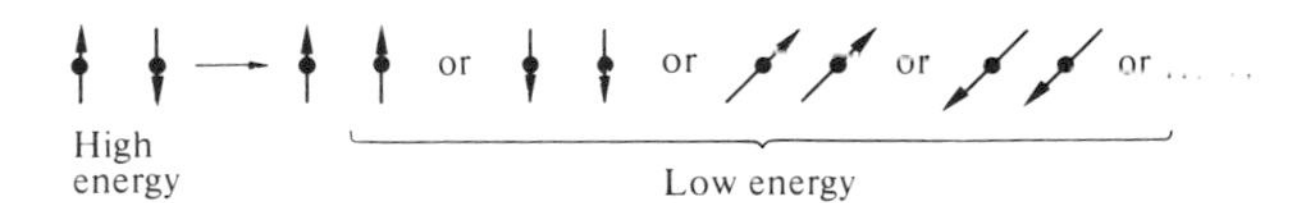

FIG. 3.12. Another example of an unstable symmetric state; one, or other, of the atomic magnets will 'flip'.

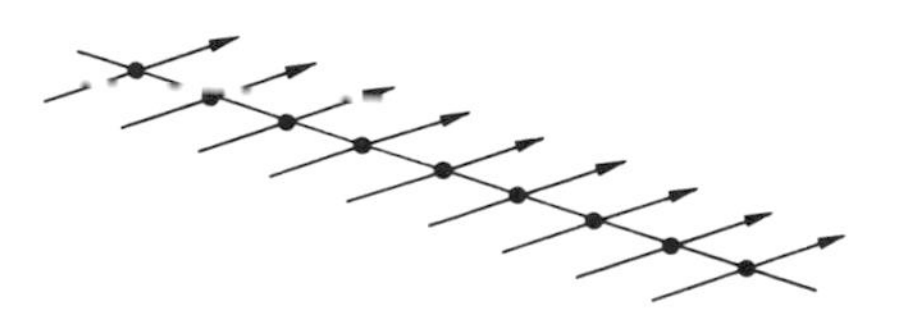

FIG. 3.13. A system of many atomic magnets can become aligned; to a 'magnetic microbe' the local environment would not appear to have rotational symmetry.

microbe living inside a magnet, it would certainly be very hard for you to detect the fact that the laws which determine the properties of the magnet are rotationally invariant.

It is easy to see that symmetries can be hidden inside a medium which can become magnetized or polarized. We live in a medium, the vacuum, which need not show the symmetries of the underlying laws. However, I must stress that, unlike the magnet, the vacuum does respect rotational invariance, it is internal symmetries which it hides.

Given enough energy, it is possible to create an electron and an anti-electron, or a quark and an antiquark, whose net attributes are zero. The Feynman diagram in Fig. 3.14(a) indicates energy, say electromagnetic energy, being converted into an electron and positron at a certain time and place. The lines indicate the electron and positron travelling away from the point at which they are created. Now, according to quantum mechanics, it takes time to test energy conservation. In fact on a very short time-scale energy need not be conserved; this is a consequence of Heisenberg's uncertainty principle for which Sir Rudolf Peierls has drawn an analogy with a

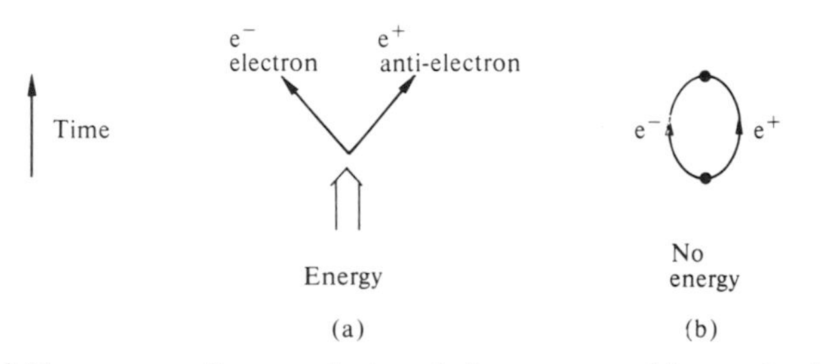

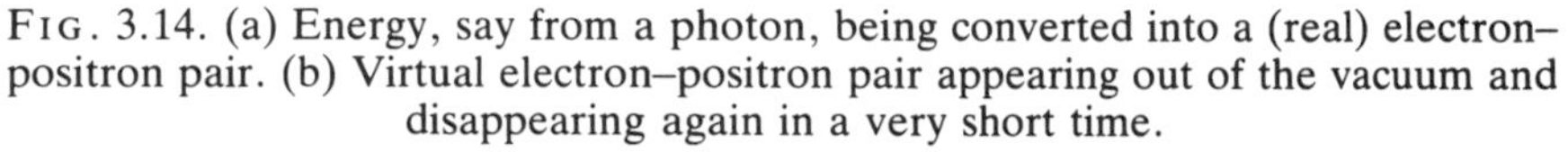

FIG. 3.14. (a) Energy, say from a photon, being converted into a (real) electron–positron pair. (b) Virtual electron–positron pair appearing out of the vacuum and disappearing again in a very short time.

dishonest cashier who borrows money: the more he borrows the sooner he must return it in order to avoid being caught. Similarly, the larger the violation of energy conservation, the shorter the time it can last. Thus it is possible to create an electron and a positron without providing any energy, but they must quickly annihilate each other (Fig. 3.14(b)). In fact such 'virtual' processes, in which pairs of particles pop out of the vacuum and promptly vanish again, are happening all the time. In their short lifetime these particles can interact, by exchanging particles, etc., and the vacuum is seething with such activity as indicated in Fig. 3.15.

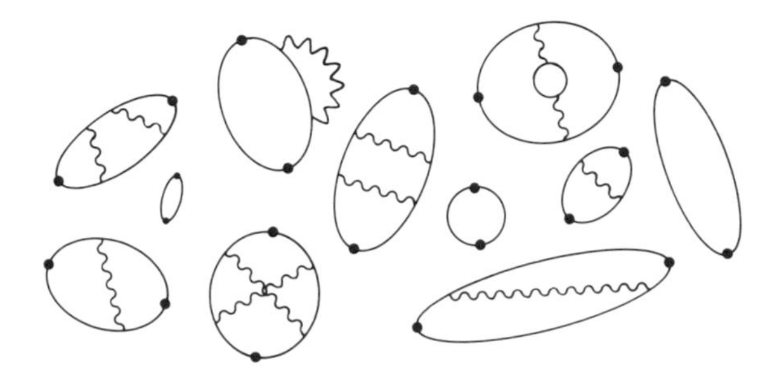

FIG. 3.15. The vacuum, seething with virtual activity.

The seething vacuum produces various effects. For example a particle travelling through it encounters 'resistance' and its mass is changed (or 'renormalized') relative to the value it would have in an empty or 'bare' vacuum. This is not necessarily a very useful piece of information since, generally, we do not know what value it would have had in an empty vacuum. However, the activities of the vacuum do have an observable effect on the force law between particles. We believe that forces are due to the exchange of particles and in chapter 1 Sir Denys Wilkinson has drawn an analogy with two skaters on a frozen lake throwing a cricket ball from one to the other. When they throw and receive the cricket ball they recoil and this gives rise, effectively, to a repulsive force between them. Similarly we believe that the electric force between a proton and an electron is due to the exchange of a photon or quantum of light (Fig. 3.16(a)). The exchanged photon can interact with the vacuum (Fig. 3.16(b)) and this gives rise to a

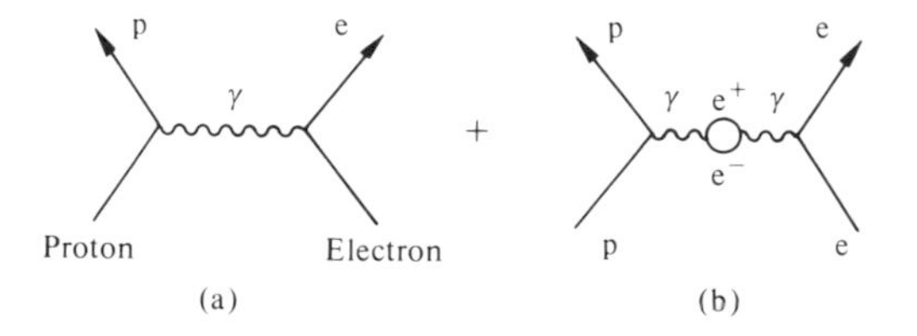

FIG. 3.16. (a) Electromagnetic interaction between proton and electron by virtual photon exchange. (b) Another process contributing to the interaction involves positron–electron pairs out of the vacuum.

small, calculable change in the force between an electron and proton as a function of their separation. This change has been checked very accurately in a variety of experiments.

For our purposes, the most important consequence of this seething activity is that the vacuum need not have the symmetries of the underlying laws. For example, the underlying laws could be left–right symmetric but the symmetric solution could be unstable. In this case the vacuum would favour left over right or vice-versa (the favoured handedness being chosen at random, like the ball falling into one valley or the other) and this could be the origin of the lack of mirror symmetry in weak phenomena.

There are two generally accepted examples of hidden symmetries in particle physics. The first is known as chiral symmetry. In this case enough relevant data are available to illustrate the fact that even when a symmetry is hidden it leads to very definite predictions which can be confronted with experimental results. To appreciate this point it is not necessary to understand the nature of chiral symmetry. Actually it involves internal symmetry operations which can turn u quarks into d quarks, and vice-versa, but which act independently on 'left-handed' quarks, which spin anticlockwise about their direction of motion, and 'right-handed' quarks, which spin clockwise (hence the name chiral from the Greek for handedness). If chiral symmetry were manifest, it would mean that left-handed and right-handed particles could never interact with each other and this would lead to many wrong predictions, for example, that the proton is massless. Hidden chiral symmetry also leads to many predictions because there is a related conservation law, although its form is rather abstruse. The predictions which have been tested are shown in the accompanying table. The meanings of the symbols need not be understood to appreciate the good, although not exact, agreement between the predicted and measured values, which constitutes clear experimental evidence for a hidden, but approximate, symmetry. The discovery that the underlying laws must exhibit (approximate) chiral symmetry

Table 3.1
Chiral symmetry

Quantity	*Prediction (Exact hidden chiral symmetry)*	*Experiment*
$\frac{M_\pi^2}{M_\varrho^2}$	0	0.03
$\frac{2M_p g_A}{f_\pi g_N}$	1	0.92 ± 0.02
$a_{1/2}$	0.16	0.17 ± 0.005
$a_{3/2}$	−0.079	−0.079 ± 0.008
λ	0.020	0.020 ± 0.003
Γ_{π^0}	7.87 eV.	7.95 ± 0.55 eV.

puts severe constraints on the possible character of the force between quarks. In fact it requires the exchanged particles which give rise to the force to carry one unit of spin. This was one of the vital clues in the discovery of the nature of the force between the quarks.

The other accepted example of a hidden symmetry is that between the weak and electromagnetic forces. Like all forces, these are due to the exchange of particles, as indicated in Fig. 3.17. Manifest symmetry between the electromagnetic and weak force would mean two things:

1. That the mass of the photon (γ) and the particles that mediate the weak force ($W^{\pm}$, W^{0}) are the same. In fact the particular form of the symmetry would require them all to be massless. Incidentally, this symmetry is called a 'gauge symmetry' and the associated conservation law includes as a special case the conservation of electric charge. Although most of the electro-weak gauge symmetry is hidden, the part called 'electromagnetic gauge symmetry', which corresponds to electric charge conservation, is manifest and this is directly connected with the masslessness of the photon.
2. That all these exchange particles are 'coupled' or tied on to the particles between which they are exchanged with the same strength. To return to the analogy of ice skaters, this would correspond to each of the skaters being able to throw each of several types of ball, $W^{\pm}$, W^{0}, and γ, equally hard.

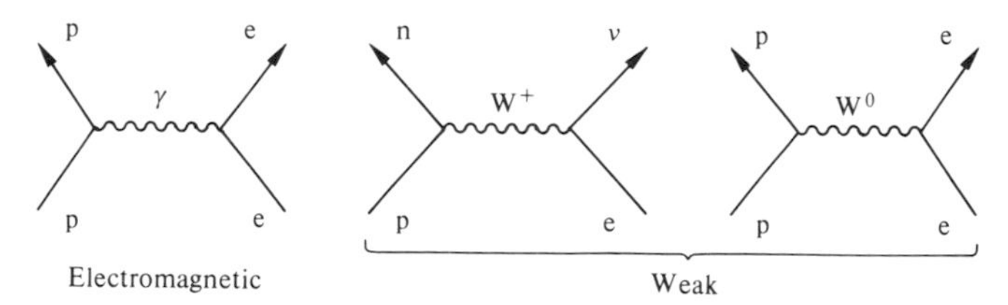

FIG. 3.17. The electromagnetic (photon exchange) and weak ($W^{\pm}$ and W^{0} exchange) interactions between proton and electron.

The properties of the vacuum do not affect the coupling of the exchanged particles which remains the same if the symmetry is hidden. However, an asymmetrical vacuum can affect their masses. In propagating through an asymmetrical vacuum the $W^{\pm}$ and W^{0} may acquire mass while the photon remains massless. It is possible to construct explicit models where exactly this occurs. The fact that the masses become different means that the symmetry will be hidden. In particular, the weak force will appear to have a very short range compared to electromagnetism since the heavier the exchanged particle, the more energy must be borrowed, the less time it can be borrowed for and the shorter the range of the interaction. A very rough analogy is provided by the ice skaters: it is clear that they can interact at a much larger distance, say 30 or 40 yards, if they exchange a cricket ball, than if they exchange a cannon ball, in which case they would have to be very close together indeed.

The effect of an interaction is the product of its strength and the volume over which it acts. The W particles having become heavy because the vacuum is asymmetrical, it follows that the effects of the resulting short range interaction due to them, the weak force, will indeed be very feeble compared to those of electromagnetism. Once the mass of the W particles has been measured we will be able to deduce the strength of their couplings and so check the symmetry with electromagnetism. This has not yet been done. Running the argument backwards, assuming the symmetry, we can predict that the mass of the W particle is almost a hundred times that of the proton. Facilities with enough energy to produce the W's are not yet available but soon will be. Meanwhile, there is a great deal of circumstantial evidence to suggest that there is indeed a hidden weak–electromagnetic symmetry.

In experiments at energies which are very large compared to the rest energy ($M_{w}c^2$) of the W particle, this symmetry would become manifest. Sufficiently energetic ice skaters could hurl light or heavy particles over equally large distances; similarly, at very high energies the mass of the exchange particle becomes irrelevant. The test for hidden symmetries of this sort is therefore either to deduce the couplings and see whether they satisfy the relations required by the symmetry, or to go to very high energies and watch the symmetry become manifest. The symmetry may also have been manifest in the very early Universe. Consider the ball which rolls into one of the two symmetrical valleys. If they are heated sufficiently, the ball will start to jump up and down with thermal energy. If enough heat is supplied it will be able to jump from one valley to the next and the symmetry will no longer be hidden. Likewise a magnet will lose its magnetic properties if it is heated sufficiently. Just after the Big Bang the Universe was extremely hot. At that time the weak–electromagnetic symmetry would have been manifest but as the Universe cooled down the symmetrical state became unstable leading to an asymmetrical vacuum hiding the symmetry.

Conclusions

The symmetries of physical laws may either be manifest or hidden. In the former case they can usually be seen directly (remember the comparison of lithium and beryllium) or inferred from conservation laws. Examples of manifest symmetries which are believed to be exact include the symmetries under translations in space and time, rotations, and electromagnetic gauge symmetry, the symmetry corresponding to the conservation of electric charge which requires the photon to be massless.

It may be that some of the observed symmetries and conservation laws will turn out to be broken when more sensitive experiments are performed. According to some theories, this will be the fate of the law of baryon conservation. The evidence for this empirical law is that no-one has ever seen a proton decay. In fact, protons are known to be stable over periods of at least 10^{30} years. However, there is no compelling theoretical basis for this law and it may well be that protons do decay very, very slowly.

Reflection symmetry is an example of a symmetry which was once thought to be exact but is now known to be violated experimentally. Whether it is violated fundamentally or is a hidden symmetry is at present unknown. The status of the symmetry between matter and antimatter, and symmetry under reversal in time is the same. It was once thought impossible to give an absolute definition of the difference between matter and antimatter, so that it would be impossible to find out which our friend in a different galaxy was made of. Similarly, it was thought impossible to tell whether a film of a sub-microscopic process was being run forwards or backwards. However, we now know that these symmetries are violated in phenomena induced by the weak interaction.

Symmetries can play an important role even when they are not exact. As an example, remember the symmetry between u, d, and s quarks which is broken by the fact that the s quark is heavier than the others. This is manifest in the approximate SU(3) symmetry of the observed hadrons which led to the invention of the quark model.

The vacuum may not exhibit some of the symmetries of the underlying physical laws; these symmetries will be hidden but nevertheless they will give rise to very definite predictions which can be tested experimentally. There is good experimental evidence for hidden (approximate) chiral symmetry and strong indications that there is a hidden symmetry between weak and electromagnetic forces. Finally, it is possible that there is a hidden symmetry between strong, weak, and electromagnetic interactions and also between quarks and leptons. Such a symmetry would have to be well hidden indeed. Strong, weak, and electromagnetic forces seem quite different and quarks appear to have nothing in common with leptons. Quarks feel the strong force, indeed the interactions between them seem to be so strong that it is impossible to separate one quark from the others to which it is bound in a hadron. Electrons, on the other hand, do not feel the strong force at all. Nevertheless it is possible to construct models in which strong, weak, and electromagnetic forces are fundamentally the same and quarks are fundamentally the same as electrons. In this case, quarks would very occasionally be transmuted into electrons and baryon number would not be conserved. Elaborate experiments are now being constructed to search for proton decay and test this idea.

Before long we may have a description of nature in which there is just one type of particle and force, the apparent diversity of particles and forces which we see being due to the fundamental symmetries being hidden.

4

Inside the Proton

D. H. PERKINS

My story starts almost 50 years ago. At that time, the Universe was supposed to be composed of a small number of varieties of fundamental constituents called, perhaps more in hope than conviction, elementary particles. These were the protons and neutrons, constituting atomic nuclei; the electrons which together with nuclei make up atoms and molecules; and the neutrino, postulated by Pauli in 1933 in connection with nuclear β-decay. Because the neutrino partakes only in the weak interactions it is very elusive and feebly interacting, and Pauli in fact predicted that the reactions of neutrinos with matter would not be detected in his lifetime. Never the less, such reactions were observed in 1955, three years before Pauli's death, at a nuclear reactor, a copious source of neutrinos.

Protons, neutrons, and electrons are commonplace, but neutrinos also make a substantial contribution to the energy content of the Universe. For example, the Sun obtains its energy from nuclear fusion reactions converting hydrogen to helium. The first stage of such an energy cycle is the conversion of two protons to a deuteron (a combination of proton and neutron) plus a positron and neutrino. Thus this is a weak interaction, and ensures that the Sun burns slowly and steadily, rather than exploding. The flux of solar neutrinos on Earth is enormous, roughly 100 million million per person per second. Because of their feeble interaction, they pass through the Earth without any absorption, and it has only been over the last few years that experiments have convincingly detected solar neutrinos.

In addition to the proton, neutron, electron, and neutrino, one should perhaps add the photon, the quantum or carrier of the electromagnetic field, to our list of 'elementary particles'. These five types of particle indeed seem to account for the vast bulk of all matter and energy in the Universe.

This simple picture of the mid-1930s began to change with the observation, just before and particularly after World War 2, of new types of particle occurring naturally in the cosmic radiation. The first new particle to be found was the muon, a sort of heavy electron (about 200 times as massive). High-energy muons constitute most of the cosmic radiation at sea-level and underground, and are of secondary origin. The primary cosmic rays are in fact high-energy nuclei—ranging from hydrogen to uranium nuclei—which are presumably ejected from stellar outbursts and may also be accelerated in galactic magnetic fields. Some of them acquire enormously high energies. The cosmic rays wander around the Galaxy, spiralling around the magnetic field lines and ultimately, after an average journey time of 2 million years, fragmenting to lighter nuclei and losing energy in collisions with interstellar matter. A few of them impinge on the Earth's atmosphere and then several things may happen. For example, if they are of low energy, they can, as a result of collisions, become trapped by the Earth's magnetic field, and execute a spiral path along the field lines, shuttling to and fro from North to South magnetic pole. In their transit, they produce excitation of the atoms in the stratosphere, resulting in the phenomenon of the aurora in the polar regions. The high-energy nuclei, on the other hand, penetrate deeper into the stratosphere and undergo nuclear collisions—strong interactions—with oxygen or nitrogen nuclei. An example of the collision of an incident iron nucleus with a nucleus in photographic emulsion carried on a high-altitude balloon is shown in Fig. 4.1. The incident and struck nuclei are fragmented to neutrons and protons, but in addition, new particles—the so-called mesons—are created out of the energy available. More than 100 charged mesons are generated in this example. These new particles are unstable, and indeed decay, in the high atmosphere, by a weak decay process to muons and neutrinos. It is these daughter muons, whose only appreciable interaction with matter is through their electric charges (that is, via the electromagnetic interactions), that penetrate to sea level.

Cosmic ray interactions can be simulated at high-energy particle accelerators to produce mesons, and hence their daughter muons and neutrinos, artificially. Next to the electron, the muon is indeed the best measured and best understood particle in the Universe. The mass, for example, is known to about 1 part per million, and the magnetic moment to about 1 part per billion. In laboratory experiments, with the help of bending and focusing magnets, we can produce very intense pencil beams of muons of almost unique energy, and use them as probes to investigate the constitution of neutrons and protons, whose structure is understood much less well. But there is one great problem with the muon. We have really no idea why it is there, what purpose it serves. Certainly, muons constitute an absolutely negligible proportion of the matter in the Universe, which could, as far as we can see, function just as well if they didn't exist. So why the muon? I shall come back to this towards the end of the chapter.

Quark patterns

The muon was only the first of many new particles found in cosmic rays

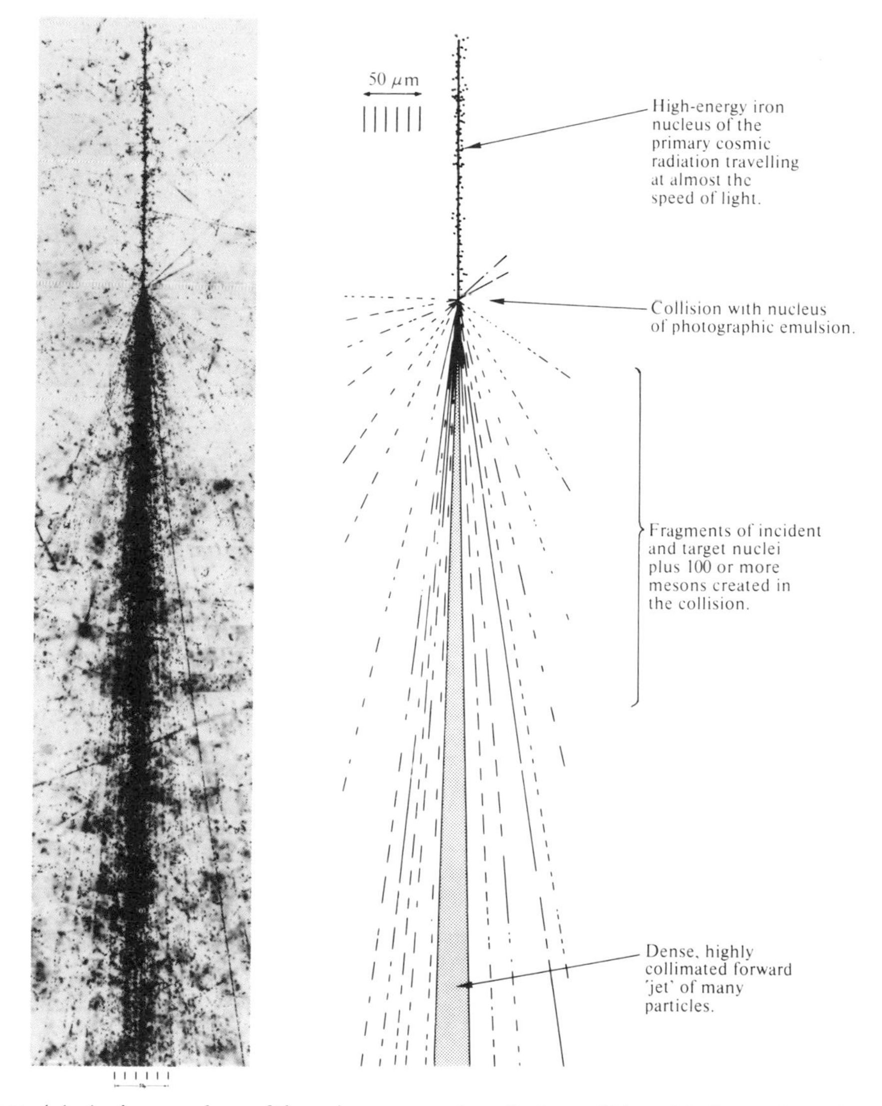

FIG. 4.1. An iron nucleus of the primary cosmic radiation collides with the nucleus of an atom in photographic emulsion carried to the top of the atmosphere by a balloon. Both nuclei fragment and 100 or more charged mesons are created.

and then at large proton accelerators. Fig. 4.2 shows the mass spectrum of the so-called baryon states that have been catalogued to date. These baryons, or heavy particles, are all very unstable and decay to protons and neutrons as end products. Most of them decay via a strong interaction (lifetime $\sim 10^{-23}$ s); a few decay via the weak interaction (lifetime $\sim 10^{-10}$ s). Apart from their instability, they appear however to be no

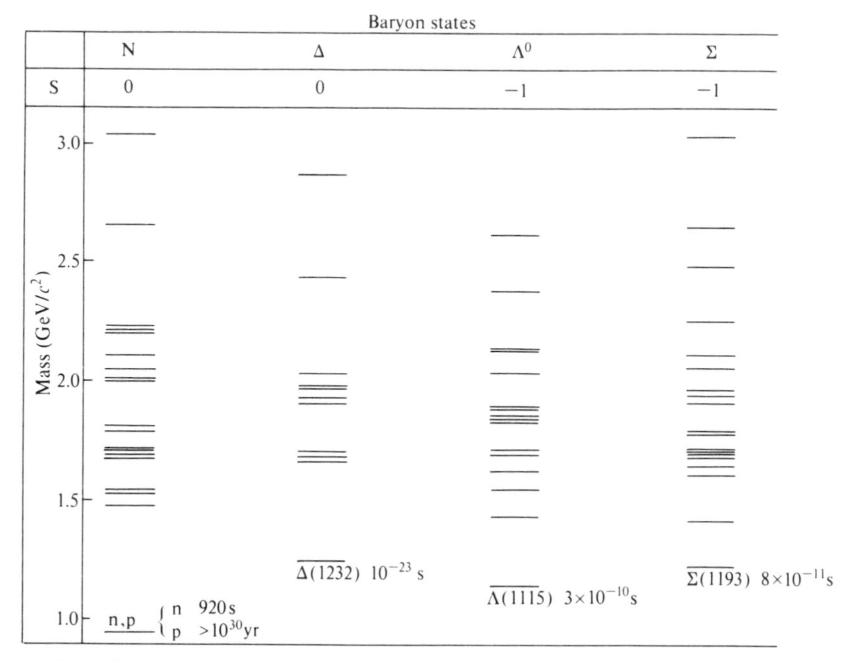

FIG. 4.2. The masses of four different types of baryon. The N series are all doublets (the lowest masses are those of the neutron (n) and proton (p) pair). The Δs are all quadruplets in electric charge (−1, 0, +1, and +2, in units of the proton charge). The Λ are all neutral singlets and the Σs are triplets (−1, 0, +1). The lifetimes of the lowest states are shown; the higher ones decay by the strong interaction with lifetimes $\sim10^{-23}$ seconds.

less or more fundamental than the proton and neutron (and even the free neutron only lives on average 15 minutes). It is possible, as explained by Chris Llewellyn Smith in chapter 3, to arrange these hundred-odd particle states into families or multiplets (characterized by a particular value of spin angular momentum). The proton and neutron fit into a pattern of eight states but occupy quite undistinguished positions in it (see Fig. 3.9(c)). This democracy of baryon states is to be contrasted again with the complete unimportance of all except the proton and neutron in the present-day Universe.

Study of the many baryon or meson states made at particle accelerators has however been a crucial step to our understanding of the proton structure. It is possible to account for virtually all the states observed simply as combinations of fractionally-charged particles called quarks (q). The list of quarks, which all have spin 1/2 (in units of Planck's constant $h/2\pi$), is given in Table 4.1. They have charges +2/3 or −1/3 (in units of e, the magnitude of the charge on an electron) and they are distinguished by different attributes, or *flavours*, with odd names: u for 'up', d for 'down', s for 'strange', c for 'charmed', etc. Antiquarks ($\bar{q}$) have opposite charge to the quarks, just as

Table 4.1
Quarks and Leptons
The fundamental fermions

Electric charge		*Quarks*		*Electric charge*		*Antiquarks*	
$+2/3$	u,	c,	?	$-2/3$	$\bar{u}$,	$\bar{c}$,	?
$-1/3$	d,	s,	b	$+1/3$	$\bar{d}$,	$\bar{s}$,	$\bar{b}$
		Leptons				*Antileptons*	
-1	e^-,	μ^-,	τ^-	$+1$	e^+,	μ^+,	τ^+,
0	ν_e,	ν_μ,	ν_τ,	0	$\bar{\nu}_e$,	$\bar{\nu}_\mu$,	$\bar{\nu}_\tau$

the antiparticle of the negatively charged electron (e^-), the positron (c^+), has an equal but positive charge.

The rule for building the strongly interacting particles or hadrons, that is the baryons (of half integral spin) and the mesons (of integral spin), is a very simple one:

$$\text{baryon} \equiv qqq,$$
$$\text{meson} \equiv q\bar{q} \; .$$

That is, a baryon is built from any three quarks of any flavours. For example, the proton is (uud) and the neutron (ddu). Even from the u, d, and s quarks alone, it is evident that 3^3 or 27 qqq combinations of flavour are possible. Furthermore, baryons of spin as high as 13/2 are known, corresponding to large values of orbital angular momentum of the quark configuration, somewhat analogous to the highly excited states of a hydrogen atom. Thus it is easily possible to account for baryon states running into the hundreds. Fig. 4.3 shows the most famous baryon multiplet, the decuplet of spin 3/2; the completion of this set of states with the discovery of the strangeness -3 Ω^- was a landmark in the establishment of the symmetry scheme for hadrons which developed into the quark model. The decuplet consists of a charge quadruplet Δ, of strangeness $S = 0$, a charge triplet of $S = -1$, a doublet of $S = -2$ and a singlet of $S = -3$. The masses of the different members of each charge multiplet are very nearly equal. Note the approximate constancy of the change in mass (~150 MeV) for each increment in strangeness. This is simply understood by writing down the quark assignments, as shown. Thus the Δ states with $S = 0$, have to be built from u and d quarks. The Σ states ($S = -1$) are obtained by exchanging a u quark for an s quark, the Ξ states by exchanging two u quarks for s quarks, and the Ω^- by changing all the u quarks in the top row to s quarks. The constant mass difference can be simply interpreted as the difference ($M_s - M_{u,\,d}$) between s and u,d quark masses. (The u and d quarks have approximately equal masses.)

The quark model concept, as first conceived by Gell-Mann and by Zweig in 1964, relied until about 1968 entirely on the above regularities in the

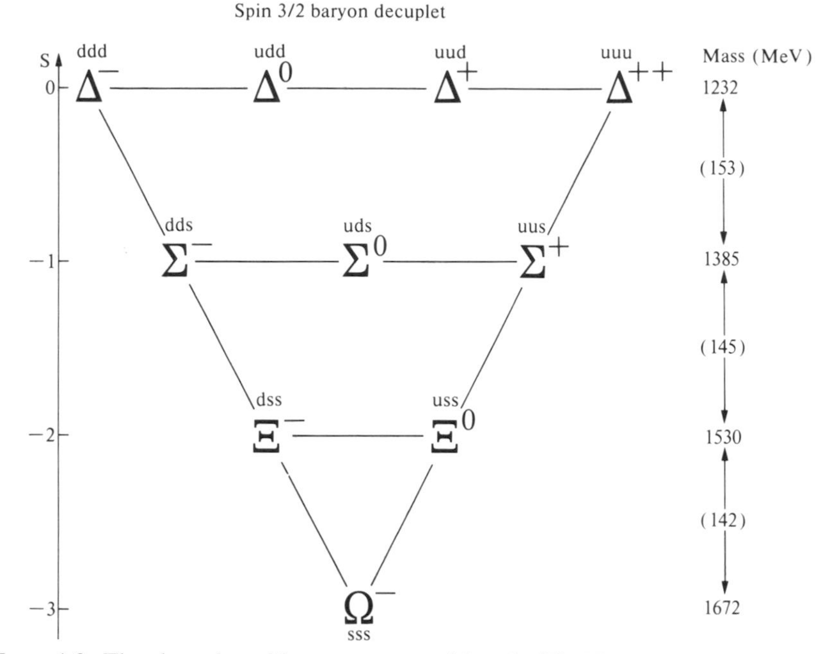

FIG. 4.3. The decuplet of baryon states with spin 3/2. The quark composition is shown for each member.

hadron mass spectra. There were, however, several difficulties inherent in the model, to set against its outstanding successes. Chief among these were the apparent absence of free quarks and serious difficulties associated with the Pauli exclusion principle.

When we say that a nucleus consists of neutrons and protons, the truth of this assertion can be proved by hitting a nucleus hard (for example with a second nucleus) and observing the neutrons and protons which come out (for example, in the collision in Fig. 4.1). A large number (more than 50) of careful and ingenious experiments have been carried out with high-energy accelerators to try to liberate the quarks. Free quarks, if produced, could hardly be missed, since their fractional electric charges would lead to very easily detectable effects—for example, they would ionize atoms in their paths at 4/9 or 1/9 of the rate of singly-charged particles. All attempts to find quarks by these means have failed. Their effects have also been searched for under a wide variety of conditions, for example in stellar spectra and in lunar rock samples. Scientists have recovered samples of deep ocean sediments (on the grounds that quarks, if heavy, would sink to these depths). They have even examined oyster shells—all to no avail. The limits on the concentrations of free quarks are really tiny. For example, in sea-water there is less than one free quark per ton, that is, per 10^{30} bound quarks. So, while no one can say that free quarks do not exist, they are certainly very rare in comparison with the ones we postulate as bound in ordinary matter.

A second difficulty arises from considering the state Δ^{++} in Fig. 4.3. The members of the decuplet have spin 3/2 and all other known spin 3/2 baryons have greater masses. Thus, we are confident that the decuplet states must be formed from quark combinations with the lowest possible energy and so with zero angular momentum generated by their relative motion. The angular momentum of 3/2 must therefore arise entirely from the intrinsic spins of the three quarks, by having the three quarks spins parallel. Symbolically we can denote the Δ^{++} state by (u $\uparrow$ u $\uparrow$ u $\uparrow$) where the arrows stand for spin vectors. We end up therefore with three identical spin 1/2 particles in the same quantum state. But the Pauli exclusion principle tells us that no two (never mind three) identical spin 1/2 particles can occupy the same quantum state. If this were not so, all of physics, chemistry, and biology would be completely different.

Probing the proton

The absence of free quarks and the violation of the Pauli exclusion principle were so serious as to suggest that quarks might be just a convenient mathematical fiction; the regularities in hadron spectra could be understood simply *as if* hadrons were built from quarks, but did not require the actual existence of such bizarre particles. This situation has been changed dramatically by experiments with lepton (electron, muon, neutrino) beams, starting about 10 years ago, in which the leptons are used to probe the internal structure of protons and neutrons.

First, we have to discuss briefly what leptons are. They are spin 1/2 point-like particles, like the quarks. Quarks and leptons are considered to be *the* fundamental fermions, or half-integer spin particles, in nature. Originally, the name meant 'light particles', and indeed the electron (e) and muon (μ) are lighter than any hadrons, but the τ-lepton has about twice the proton mass. Charged leptons have a charge of -1 and the charged antileptons have, as in the case of the quarks, the opposite electric charges. The neutral leptons are the neutrinos. They have, as far as we know, practically zero mass. Each charged lepton has accompanying it a distinctive neutrino as indicated by the subscripts. Thus, in the β-decay of a pion, $\pi^+ \rightarrow \mu^+ + \nu_\mu$, the neutrino is a muon-type neutrino and in subsequent interactions will always produce a charged μ, not an e or τ; thus, $\nu_\mu + N \rightarrow P + \mu^-$, for example. The neutral antileptons are called antineutrinos. They are distinct from neutrinos in respect of their spin polarization: we can think of neutrinos as having spins, σ, pointing against their momentum vectors P (i.e. defining a 'left-handed' screw system) while antineutrinos are 'right-handed'; as illustrated in Fig. 4.4.

Leptons are distinguished from quarks in their interactions with other particles. Neutrinos only have weak interactions, and charged leptons only electromagnetic and weak interactions. Quarks on the other hand have not only electromagnetic and weak interactions with other quarks and with

FIG. 4.4. Diagram illustrating the difference between 'left-handed' and 'right-handed' correlations of spin and motion. The direction of spin is conventionally indicated by an arrow pointing in the direction along which the spin appears clockwise. If this arrow (or vector σ) is opposite to the direction of the particle's motion (momentum vector P) this is like a 'left-handed corkscrew'; the opposite situation is 'right-handed'. Neutrinos are left-handed, anti-neutrinos are right-handed.

leptons, but in addition experience the much more powerful strong interactions between themselves.

I shall now discuss briefly, by way of illustration, the production of neutrino (or antineutrino) beams at accelerators, before describing the experiments. The method employed mirrors almost exactly what happens naturally when cosmic ray primaries interact in the Earth's atmosphere. First, a proton synchroton (see chapter 7) is used to accelerate protons to high energy. The CERN super proton synchrotron, for example, consists of a ring of bending and focusing electromagnetics placed in an underground tunnel 7 km in circumference. The protons, about 10^{13} of them, circulate in a vacuum pipe threading the magnets. They are accelerated a few times per revolution in a radio-frequency electric field and constrained in their circular path by the ring of magnets. After about 0.5 million revolutions of the ring, taking nearly 2 seconds, they have achieved the peak energy of 400 GeV when they are extracted from the machine and shot into a beryllium target. In the violent nuclear collisions which follow, large numbers of mesons are produced, just as in the cosmic-ray event of Fig. 4.1. The beam layout is shown in Fig. 4.5. A long line of focusing and bending magnets and collimators serves to select mesons (mostly pions) of one sign of charge and a particular momentum. These enter an evacuated decay tunnel, where a fraction of them decay in flight into muons and neutrinos, the analogue of atmospheric decay in the cosmic-ray process. Finally, a steel and rock shield, some 350 m thick, is needed first to absorb the remaining mesons, mainly through dissipation by nuclear collision in the first metre or so, and then to stop the muons through a very gradual sapping of their energy in electromagnetic encounters with the atoms of iron and rock. This leaves a pure beam of neutrinos if positive pions are selected ($\pi^+ \rightarrow \mu^+ + \nu_\mu$), or antineutrinos if the pions are negative ($\pi^- \rightarrow \mu^- + \bar{\nu}_\mu$).

The acceleration cycle takes about 8 seconds, so with this periodicity an intense burst of neutrinos emerges from the end of the shield and traverses several detectors placed in series. The first of these is a large bubble chamber

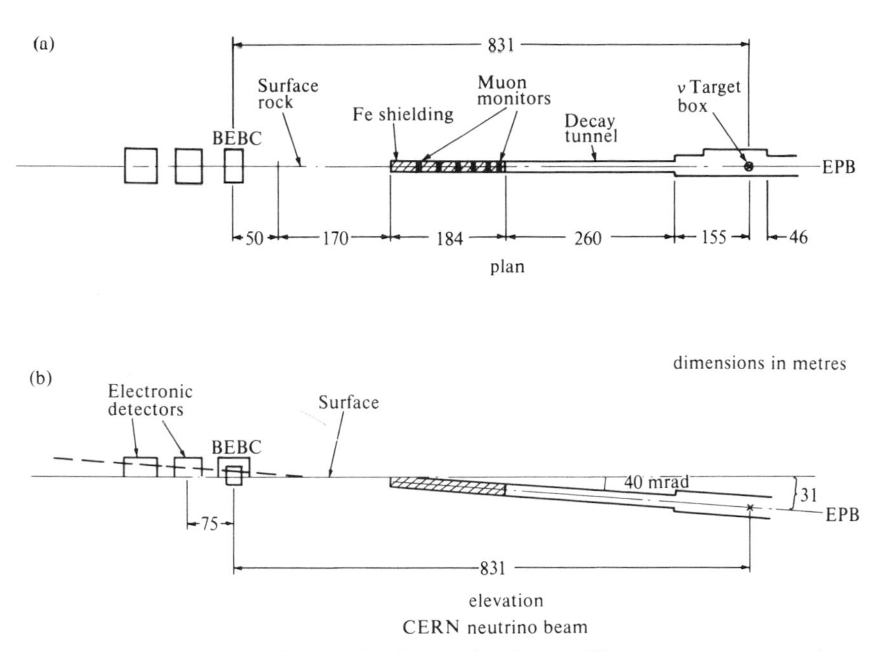

FIG. 4.5. Layout of the CERN SPS neutrino beam. The extracted proton beam (EPB) hits a beryllium target, generating mesons which are focused into a beam pointing towards the detectors about 1 km away. Decay to muons and neutrinos occurs as the mesons pass along the decay tunnel. The remaining mesons and the muons are removed by the iron and rock to leave the neutrinos.

BEBC (Fig. 4.6), containing 10 tons of neon–hydrogen mixture. The liquid is maintained in a pressure vessel in a superheated state, but prevented from boiling by an overpressure. A few milliseconds before the neutrino burst reaches the chamber, the pressure is released and the liquid tends to boil. Initially the boiling takes place along the trails of electrons and positive ions (atoms which have lost an electron) made in the liquid by the fast charged particles generated by a neutrino interaction (the neutrino, like other uncharged particles, cannot knock electrons out of atoms and so does not make a trail of bubbles). These strings of bubbles are then photographed, before the liquid is re-compressed to stop boiling and prepare for the next burst of neutrinos. The momentum of a particle may be determined from the track curvature in the applied magnetic field (about 35 kgauss, from a superconducting magnet). Fig. 4.7 shows an electronic detector, of total mass about 1500 tons. It consists of magnetized iron plates interleaved with scintillation counters and drift chambers which allow particle trajectories to be reconstructed from the co-ordinates, recorded on magnetic tape, of the points 'hit' by the particles as they pass through them. Roughly, 10^9 neutrinos traverse this massive detector in each burst, but only a handful of them interact.

FIG. 4.6. BEBC, the Big European Bubble Chamber. Nearly 4 m in diameter the chamber can be filled with 10 tons of liquid neon–hydrogen mixture, or about 1 ton of liquid hydrogen. It is surrounded by superconducting coils of niobium–titanium which can generate a magnetic field of 35 kilogauss. In this view can be seen the planes of electronic detectors (multi-wire-proportional-chambers) placed around the exit sides to identify muons (CERN).

FIG. 4.7. Another experiment to study neutrino interactions. It consists of about 1500 tons of iron interleaved with electronic particle-detectors. A second massive detector system is also in the picture, further downstream (CERN).

Fig. 4.8 shows a photograph of a neutrino interaction in the bubble chamber. In this event, the neutrino, ν_μ, of energy ~200 GeV transforms into a negative muon (identified by external detectors) of energy 100 GeV, and the remaining energy is used to create several mesons: ν_μ + (N or P) $\rightarrow$ μ^- + hadrons. The configuration of secondary particles looks (and is) complicated, and it is far from obvious how one is going to learn anything about neutron and proton structure from it. Suppose one asks a much simpler question, however: how does the probability of interaction (that is, the cross-section presented to a neutrino by an individual proton or neutron) depend on the neutrino energy? Fig. 4.9 shows that the answer is indeed simple: the cross-section is very nearly proportional to the energy. What does this result tell us? At this point, we need a result from quantum mechanics. For elastic scattering (that is, a collision in which two particles enter and two leave) between two structureless particles the cross-section is just proportional to the density of final states in 'momentum space' which, for this process, is proportional to the energy of the incident neutrino. Thus the fact that the neutrino cross-section is observed to be proportional to the

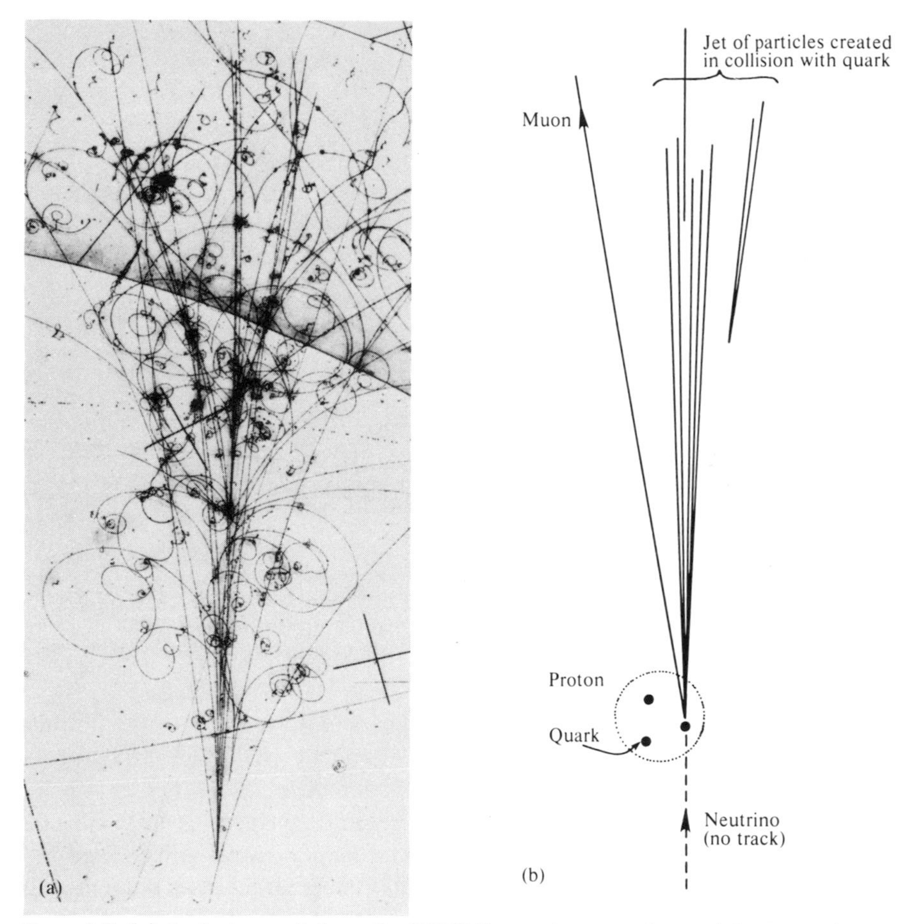

FIG. 4.8. (a) A photograph from the BEBC neutrino experiment (also reproduced on dust cover) (b) A diagram illustrating the main features of the neutrino interaction and its interpretation as a collision with a pointlike constituent (a quark) within a proton.

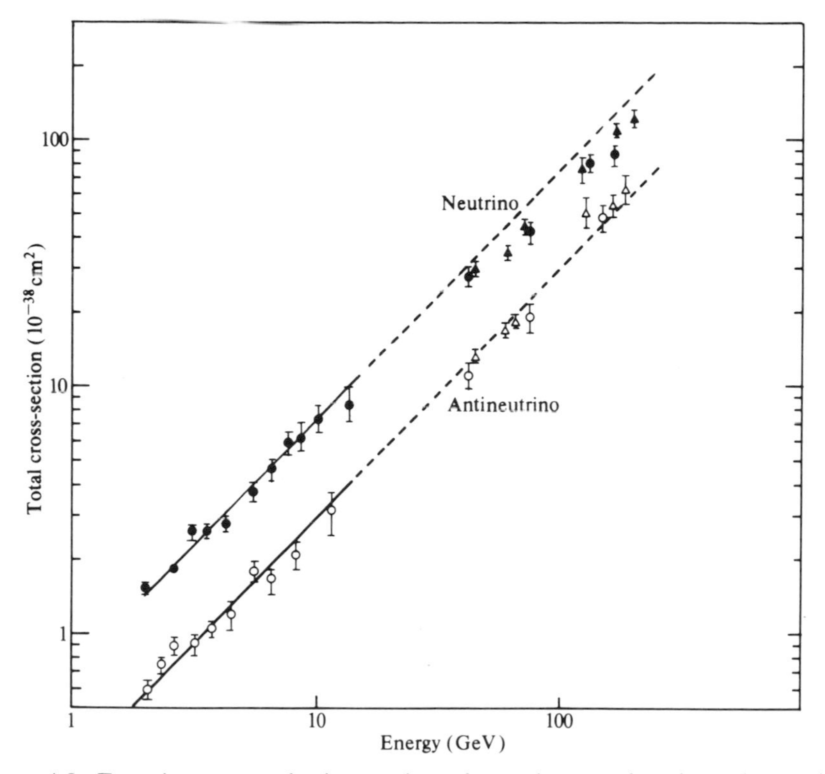

FIG. 4.9. Experiments on the interaction of neutrinos, and antineutrinos, with energies from a few GeV to over 100 GeV shows the total cross-sections rising in proportion to the neutrino energy.

neutrino energy implies in turn that complicated reactions like the one shown in Fig. 4.8 appear to be the end result of something which started as the elastic scattering of two point-like particles. It must mean that the proton (or neutron) target contains pointlike, quasi-free constituents, one of which scatters the pointlike neutrino independently of the other constituents. This situation is represented pictorially in Fig. 4.10. Here, we regard the reaction as a two-stage process. First, the neutrino scatters elastically off one of the pointlike constituents, and it is this first stage which determines the form of the cross-section. In the second stage, shown as a 'black box' to indicate that we do not know the exact mechanisms involved, the pointlike constituents must recombine to form the hadrons we observe. Our hypothesis is that the first and second stages take place independently; in no other way could we end up with an exactly linear relationship between cross-section and neutrino energy.

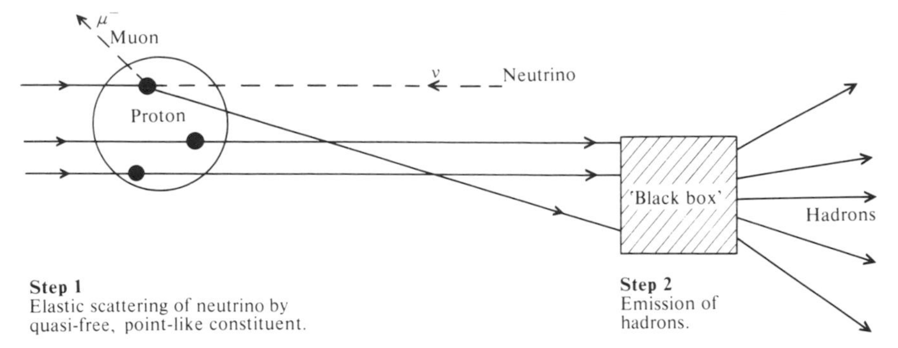

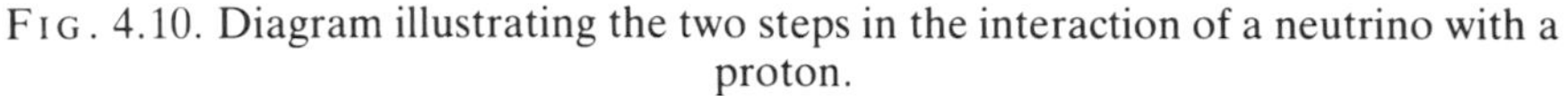
FIG. 4.10. Diagram illustrating the two steps in the interaction of a neutrino with a proton.

The constituents

Having discovered the pointlike character of the proton constituents, we have to establish their identity (they will turn out to be the quarks). First, we can measure the spin of these objects. This is done in other scattering experiments, using this time charged lepton beams, that is, electrons or muons. The experiments compare the forward scattering of electrons, in which the incident electron is almost undeflected, with the large angle, or backward scattering. Now the forward scattering is determined by the electrical interaction between charges, while the backward scattering is determined by the interaction between the magnetic moments of the particles involved. One can plot the ratio of backward to forward scattering probability in such a way as to determine the magnetic moment, or equivalently the spin of the constituents. In fact, in Fig. 4.11 the quantity given on the ordinate is $g/2$, where g is the so-called gyromagnetic ratio, which is predicted by Dirac's theory of pointlike spin 1/2 particles to be 2. If the constituents have spin 1/2 therefore, the data should indicate a value of unity for $g/2$. If the spin were zero, the magnetic scattering would vanish and the points would lie along the x-axis. The data clearly indicate spin 1/2 constituents.

Plotted horizontally in this graph (Fig. 4.11) is a quantity x which depends on the angular deflection of the electron or, more precisely, q^2 which is the square of the momentum transferred by the electron to the constituent, and also on its energy loss, ω. These quantities are related to the mass m of the constituent in the elastic scattering process:

$$q^2 = 2m\omega; \quad x = q^2/2M\omega = m/M,$$

where M is the proton mass. So x is the fraction of the proton (or neutron) mass carried by the struck constituent.

Thus if the constituent has small mass, the electron will be little deflected, and q, the momentum transfer, is small; while if the electron strikes a heavy constituent, it will be deflected through a big angle. One might wonder here

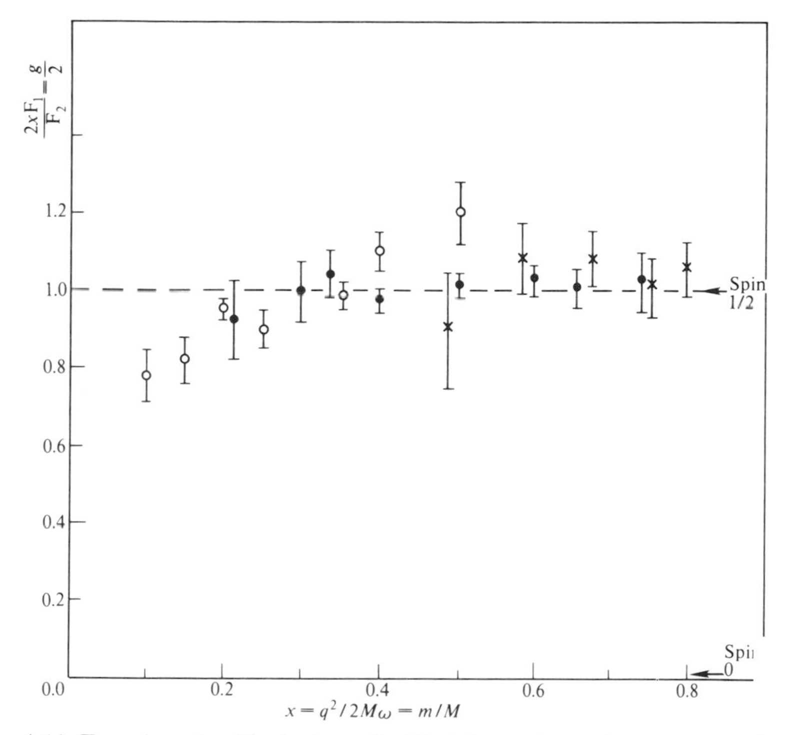

FIG. 4.11. Experiments with electrons find that the proton and neutron constituents have a value of $g/2$ close to 1, the value expected in Dirac's theory for point-like particles of spin 1/2.

why the constituents do not have a unique mass. If they are quarks, and there are three in a proton or neutron, surely democracy would demand that each has a mass m equal to one-third of the total mass, that is $x = 1/3$? However, one has to take into account that the quarks are confined inside the linear dimensions ($R \sim 10^{-13}$ cm) of a proton, and quantum mechanics then tells us that they will have a 'Fermi momentum'. This is just an application of the famous uncertainty principle of Heisenberg which, in this situation, states that it is impossible to measure simultaneously both the position, r, and the momentum, p, of a particle to arbitrary accuracy; the 'uncertainties' must obey $\Delta p. \Delta r > \hbar$. But if we *know* that $\Delta r < R$, then there must be a corresponding uncertainty in the momentum, given by $\Delta pc \sim \hbar c\ R \sim 200$ MeV. Depending on whether the quark is 'running away from' or 'towards' the incident lepton, the mass of the equivalent stationary quark is changed by a factor $\pm \Delta pc/Mx \sim 0.2/x$. Thus, this 'zero point kinetic energy' of the bound quarks would effectively smear out any peak in the fractional mass distribution.

Fig. 4.12 shows the neutrino cross-section data plotted as a function of x. The vertical scale is the cross-section divided by the incident energy (that is, taking out the dependence on the kinematic factor proportional to the

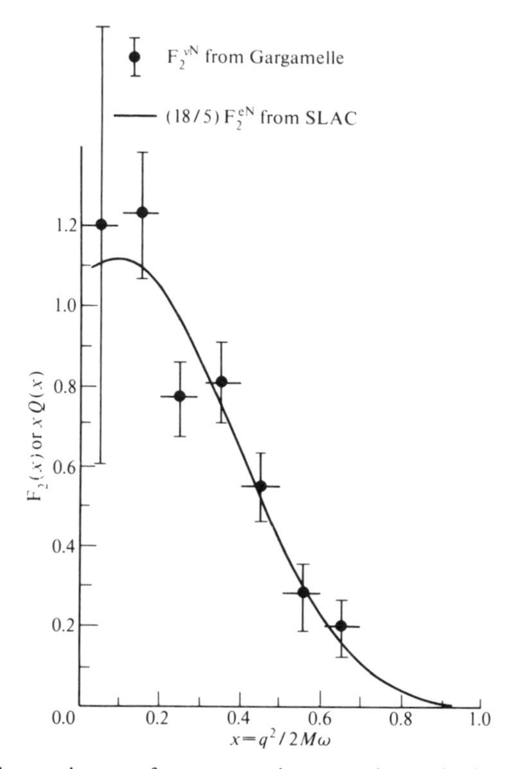

FIG. 4.12. The data points are from a neutrino experiment in the Gargamelle heavy liquid bubble chamber at CERN, while the curve is obtained by multiplying data from electron scattering experiments, performed at the Stanford linear accelerator (SLAC), by a factor 18/5. The agreement between curve and data points suggests that the constituents of protons and neutrons have the fractional charges (1/3 and 2/3) postulated for the quarks.

density of final states) and it is then a measure of the quantity $xQ(x)$, where $Q(x)$ is the probability of there being a quark with that value of x. This product, $xQ(x)$, is actually the fraction of the proton's mass carried by those constituents, quarks, having a mass $m = xM$.

A similar analysis can be made for the electron, or muon, scattering experiments. In these cases, however, because the probability of scattering through the electromagnetic interaction is proportional to the square of the electric charge of the scattering object, the cross-section at a given x depends on $xQ(x)$ multiplied by the square of the quark charge. The experimental results we are using were obtained with target nuclei containing equal numbers of protons and neutrons, thus also equal numbers of u and d quarks. The charges, squared, have the values 4/9 for a 'u' and 1/9 for a 'd', with a sum of 5/9 and so an average, or mean square charge of 5/18. Thus, if the electron data is multiplied by 18/5, it should agree with the neutrino data. Indeed it does, to within the experimental error of about 10 per cent. It is difficult to see how one can obtain such an unlikely number as 18/5 just by accident. These results are therefore confirmation that both neutrinos and electrons 'see' the same substructure in the proton, one reacting to 'weak charges' and the other to electrical charges, and that the quarks really do have those weird fractional charges.

There is however, one very disturbing feature of the plot in Fig. 4.12. We said that $xQ(x)$ was the fraction of the proton's mass carried by quarks having a fractional mass x; thus the sum of this quantity for all values of x, that is $\int xQ(x)\,dx$, which is the area under the curve of Fig. 4.12, should be a measure of the total mass fraction carried by quarks. If the proton consists only of quarks this integral clearly has to be unity. It obviously is not—it is about one half! Something is badly wrong somewhere, we have a missing mass problem.

To sum up at this point; before lepton scattering experiments got started, in the late 1960s, the quark concept rested on regularities in hadron spectra but suffered from two big problems; no free quarks had been seen and the Pauli principle was violated in construction of the Δ^{++} resonance out of three identical u-quarks required to be in the same physical state of relative motion and spin orientation. A huge effort, extending over the last 10 years, on lepton–nucleon scattering measurements, has given strong verification of the nature of the nucleon constituents as quarks but now we are faced with *four* problems rather than just two. The first is the absence of free quarks, which might be ascribed, in a way which is not presently understood, to some confinement mechanism which undoubtedly requires very *strong* binding forces between the quarks, to prevent them getting out. Second, the Pauli principle violation is still there. Third, the successful description of lepton scattering by neutrons and protons in terms of elastic scattering by quasi-free quarks equally requires very *weak* binding forces between them, apparently inconsistent with objection number one. And finally, we have discovered a missing mass problem.

The colour force

If we by-pass the confinement mystery, then, over the last five years, a theory of quark—quark interactions has been developed which clears all the last three hurdles at one bound! It is called quantum chromodynamics, or QCD for short. To understand what this theory is about, it is useful to consider first the astonishingly successful quantum field theory of electromagnetism, called quantum electrodynamics, or QED. In quantum field theory, two electric charges interact by exchange of a virtual photon (the carrier of the electromagnetic field). This photon is virtual in the sense that energy ΔE can be borrowed to create the quantum, provided it is done only for a short time Δt, where $\Delta E.\Delta t \sim \hbar$; the uncertainty principle again.

So, we can imagine an electron, for example, as continually emitting and re-absorbing virtual photons (or, they could be absorbed by another charge if there happens to be one nearby). As an example, let us consider the situation when an electron is placed in a magnetic field for an experiment to determine its magnetic moment, or gyromagnetic ratio, g, by measuring the spin precession frequency. Pictorially, this can be represented by the series of Feynman diagrams shown in Fig. 4.13, which help the physicist to calculate what happens.

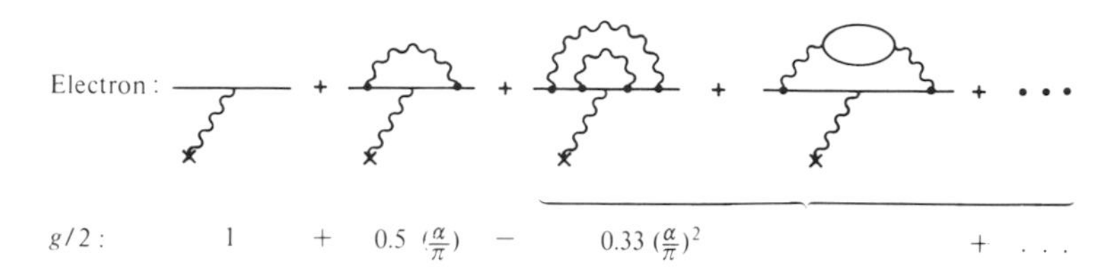

FIG. 4.13. The first few terms of the series in powers of α (the fine structure constant) determining the quantum electrodynamics (QED) corrections to the electron's gyromagnetic ratio, g.

The single horizontal line of the first 'term' depicts a 'bare' point-like electron interacting with a photon of the applied electromagnetic field; this situation is described exactly by Dirac's theory for a spin 1/2 particle which predicts the value 2 for g. (The gyromagnetic ratio is the magnetic moment expressed in units of the Bohr magneton, $e\hbar/2mc$, divided by the spin angular momentum in units of $\hbar$.) The second term shows the electron temporarily dissociated into an electron and one virtual photon; if one applies a magnetic field to an electron in this condition the magnetic moment, which depends on the ratio of charge to mass (e/m), will be bigger, because whereas all the charge (which couples to the applied field) still resides on the electron, part of its mass has been momentarily transferred to the energy of the virtual photon. The correction to g is proportional to the probability of virtual photon emission, given by the 'fine structure constant' α ($\alpha = e^2/\hbar c = 1/137$) which specifies the strength of the coupling of photons to charges. Another possibility, indicated by the two other diagrams, is the emission of two photons, or a single photon interacting with a virtual electron–positron pair out of the 'seething vacuum' described by Chris Llewellyn Smith in chapter 3; both these processes contribute to the term of order α^2. Proceeding in this way, the correction to the g-factor can be expressed as a power series in α. The labour involved in these calculations is enormous—there are a total of 72 terms of order α^3. All the α^4 terms will probably never be calculated (the number of Nth order terms goes like N^N). Furthermore, since the theory relies on the experimental value for α, there is a point at which further theoretical precision is overshadowed by the propagated error in α.

Including terms up to α^3, the theoretical prediction is:

$$\frac{g}{2} = 1.001\ 159\ 652\ 4 \pm 0.000\ 000\ 000\ 4,$$

while the experiments give:

$$\frac{g}{2} = 1.001\ 159\ 652\ 4 \pm 0.000\ 000\ 000\ 2.$$

The corresponding values for the muon are:

$$\frac{g}{2} = 1.001\ 165\ 921 \pm 0.000\ 000\ 009, \text{ theory},$$

$$\frac{g}{2} = 1.001\ 165\ 924 \pm 0.000\ 000\ 008, \text{ experiment}.$$

There is incredibly good agreement between experiment and theory. Such a statement really does not justice at all either to the theoretical effort involved or the beauty and ingenuity of experiments of ever greater precision, developed over some three decades.

In summary then, QED is an outstandingly successful theory of the interactions of charged particles and photons. Small wonder therefore that physicists should aspire to describe the quark interactions by a similar type of theory. Basically, QCD postulates specific quanta—called gluons—mediating the strong interactions between quarks, just as photons mediate interactions between electrical charges (see Fig. 4.14). There are however, fundamental differences. The strong charge is called 'colour', and it comes in three varieties, compared with just two varieties (positive and negative) for electric charge. Colour is just a name for the extra degree of freedom one has to assign to quarks in describing their strong coupling.

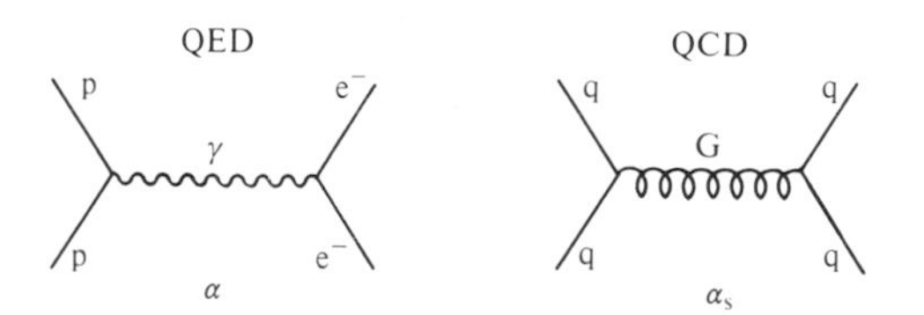

FIG. 4.14. Comparison of virtual photon exchange mediating the electromagnetic interaction (QED) between electric charges and virtual gluon exchange responsible for the force (QCD) between quarks carrying 'colour charges'. The QED coupling is specified by the fine structure constant, α, and QCD by an analogous quantity, α_s.

While the photon itself has no electric charge, the gluons do carry a colour charge, as do the quarks. This has a dramatic effect on the value of the strong coupling constant, α_s, at very small distances. We can perhaps understand this in terms of the phenomenon of 'screening' of electrical charges. If we surround a charge e by a dielectric material (in which the molecules are polarized by an applied electric field, that is their oppositely charged parts tend to move apart) the influence of the charge at large distances is reduced compared with the case of no dielectric. This is because the polarization charges shield the charge (Fig. 4.15). This effect is important at distances large compared with molecular dimensions; at small distances, one obtains the effect of the bare or unscreened charge. The effective coupling, α, of two electric charges therefore would depend on the distance separating them. Actually, this happens even without a dielectric, since if the charge e were placed in vacuum, it would be surrounded by charges due to virtual electron–positron pairs (Fig. 4.16(a)), the so-called 'vacuum polarization'. That this effectively increases α at small distances is demonstrated by the relative g values of electron and muon, given above. Since the muon is heavier, it involves virtual pair states of higher momentum, that is probing to smaller distances, than does the electron and indeed we find the effective value of α, as measured by $g/2 = 1 + 0.5(\alpha/\pi)$, is slightly bigger for the muon.

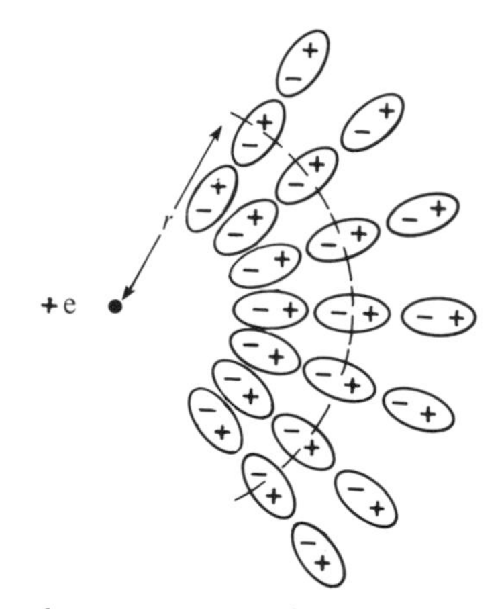

FIG. 4.15. Polarization charges screen a charge e placed in a dielectric medium.

Screening effects are also expected in QCD but lead to the opposite behaviour. Because the gluons carry the 'strong charge' of colour, whereas the photon does not carry electromagnetic charge, a new process (Fig. 4.16(c)) of virtual gluon production by gluons with no counterpart in QED occurs; the effect of this process dominates the QED-like term of Fig. 4.16(b) at small distances (large momenta). Thus the gluons can 'spread out' the colour charge so that for close, high-momentum-transfer collisions the effective coupling strength of the interaction, α_s, is reduced.

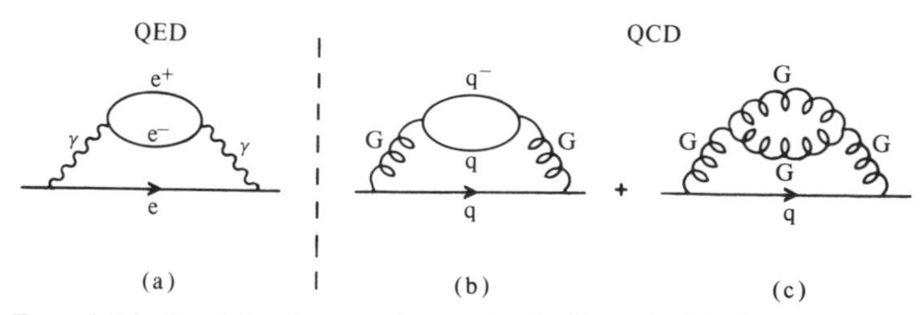

FIG. 4.16. Contributions to 'screening' effects in QED and QCD.

So, the distinctive feature of QCD is that the coupling is predicted to become weak at small enough distances, it is said to be 'asymptotically free'. This property is consistent with the empirically observed behaviour of deep inelastic lepton–nucleon scattering; the constituent quarks behave as if almost free in the high q^2, small distance region. By the same token, the coupling will become progressively stronger at large distance, and could presumably lead to confinement of quarks. Since gluons also carry the colour charge they must also remain trapped inside the hadrons. Actually, at large distance and for strong coupling ($\alpha_s \sim 1$) the theory does not give a calculable answer (the power series in α_s would not converge) and confinement is still a quantitatively unsolved problem.

Evidence for colour

What is the quantitative, rather than qualitative, evidence in favour of QCD? We have postulated colour as an extra degree of freedom. Quarks come in three colours. The best evidence in favour of this assumption is from the process of $e^+ e^-$ annihilation at high energy. Two types of process observed are:

$$e^+e^- \rightarrow \mu^+\mu^-,$$
$$e^+e^- \rightarrow \text{hadrons (e.g. pions)}.$$

For the first process, QED predicts an angular distribution of the emergent muons of the form $(1 + \cos^2\theta)$, where θ is the angle of a muon relative to the direction of the colliding e^+ and e^-, measured in the laboratory frame, where they have equal and opposite momenta (Fig. 4.17(a)).

For the process $e^+e^- \rightarrow$ hadrons, it is observed that the secondary mesons emerge in two oppositely moving 'jets'. An example is shown in Fig. 4.18. Constructing the resultant momentum vectors of these jets, the angular

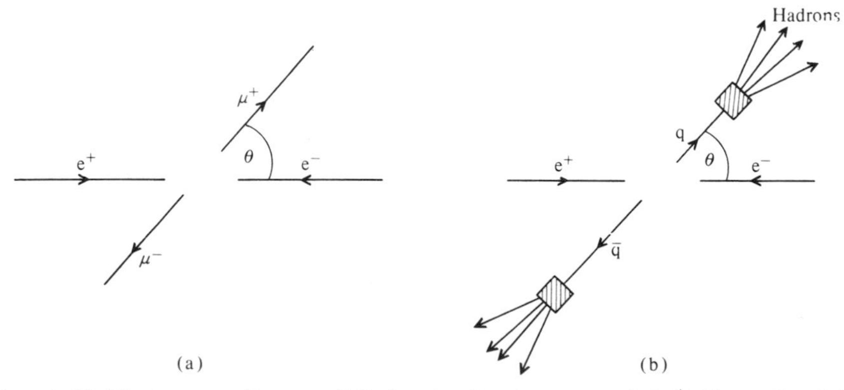

FIG. 4.17. Electron–positron annihilation leading to muon pair ($\mu^+\mu^-$) creation (a), and (b) hadron production, via quark–antiquark pair ($q\bar{q}$) creation as a first step.

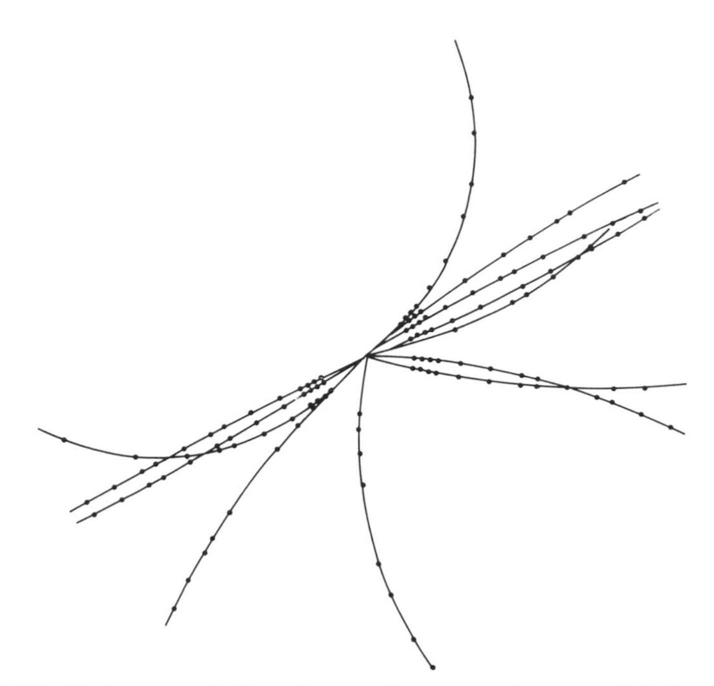

FIG. 4.18. Hadron production in an electron–positron annihilation at the PETRA machine, DESY Laboratory, Hamburg. The trajectories are reconstructed from data obtained with the TASSO detector and the view is along the direction of the colliding beams. The hadrons emerge in two, back-to-back, 'jets' and their paths are curved by a magnetic field.

distribution is again found to be of the form $(1 + \cos^2\theta)$, see Fig. 4.19. Since this distribution is typical of the production of two spin 1/2 particles in the final state, as in the case of $\mu^+\mu^-$ production, we naturally interpret the process as formation of a quark–antiquark pair followed by the 'black box' process of conversion of quarks to hadrons (Fig. 4.17(b)).

The cross-section of these electromagnetic processes involving elementary spin 1/2 particles, be they muons or quarks, must depend on energy in the same way and be proportional to the charge, squared, of the particles involved. Thus the ratio of the cross-sections for hadron and muon pair production in e^+e^- annihilation is given by

$$R = \frac{\sigma\,(e^+e^- \rightarrow q\bar{q} \rightarrow \text{hadrons})}{\sigma\,(e^+e^- \rightarrow \mu^+\mu^-)} = \frac{\Sigma\,(c_i)^2}{1},$$

where the sum, $\Sigma(c_i)^2$, is taken over all the charges, c_i, of the different types of quark that can be produced. Fig. 4.20 shows that, for collision energies below 4 GeV, where only $u\bar{u}$, $d\bar{d}$, and $s\bar{s}$ quark pairs can be created, we therefore expect $R = (2/3)^2 + (-1/3)^2 + (-1/3)^2 = 2/3$. But, if each quark flavour comes in three colours, there are three times as many different types of final states, and we should find $R = 2$. The observed R values are

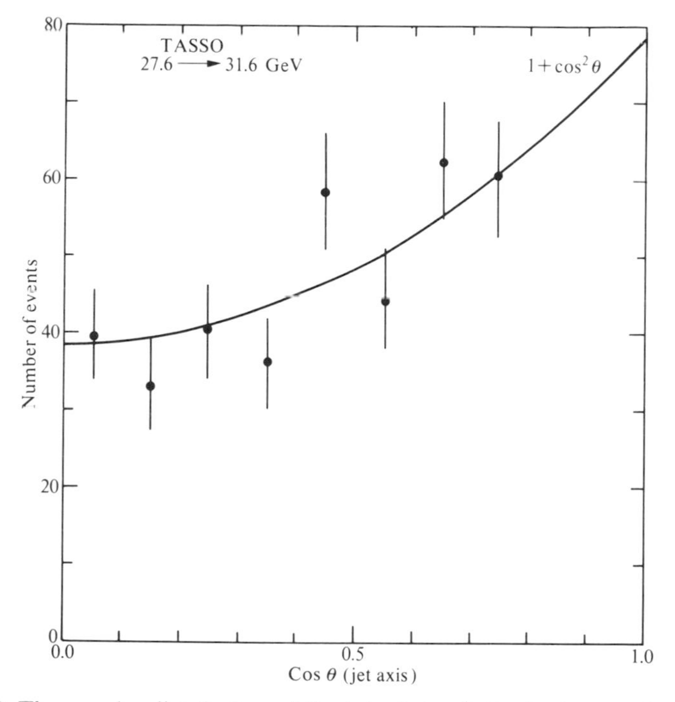

FIG. 4.19. The angular distribution of the jet axis in $e^+e^- \rightarrow$ hadrons, relative to the beam direction. The data is from the TASSO experiment.

close to (but slightly above) the value of 2. For higher energies, above 10 GeV, heavier quark pairs $c\bar{c}$ and $b\bar{b}$ can also be produced, predicting $R = (2/3)^2 + (-1/3)^2 + (-1/3)^2 + (2/3)^2 + (-1/3)^2 = 11/9$ without colour, or 11/3 with colour. Again the data are in favour of the colour scheme, although slightly high. In fact the full QCD prediction is that the predicted ratios given above should apply only at infinite energies, and that R should be somewhat (~10 per cent) higher at the energies shown. So everything agrees pretty well.

We also see that colour gets us out of our Pauli exclusion principle problem. However we do not need the colour degree of freedom to describe hadrons, we have attributes enough for them. In other words the three colours of the three quarks in a baryon must combine to form a colourless object, what we call a colour singlet, with, say, one red, one green, and one blue quark. Thus, the three u quarks in a Δ^{++} are no longer identical.

What about the gluons we have introduced? We could identify the quarks by virtue of their magnetic moments or electrical charges, but gluons possess neither. However, there are effects which might allow a gluon to be 'seen'. We expect, as in Fig. 4.21, that one of the quarks of Fig. 4.17 will occasionally radiate a high-energy gluon at large angle, and that the gluon and quark will then 'dress up' to form separate jets of hadrons. Such '3-jet events' seem

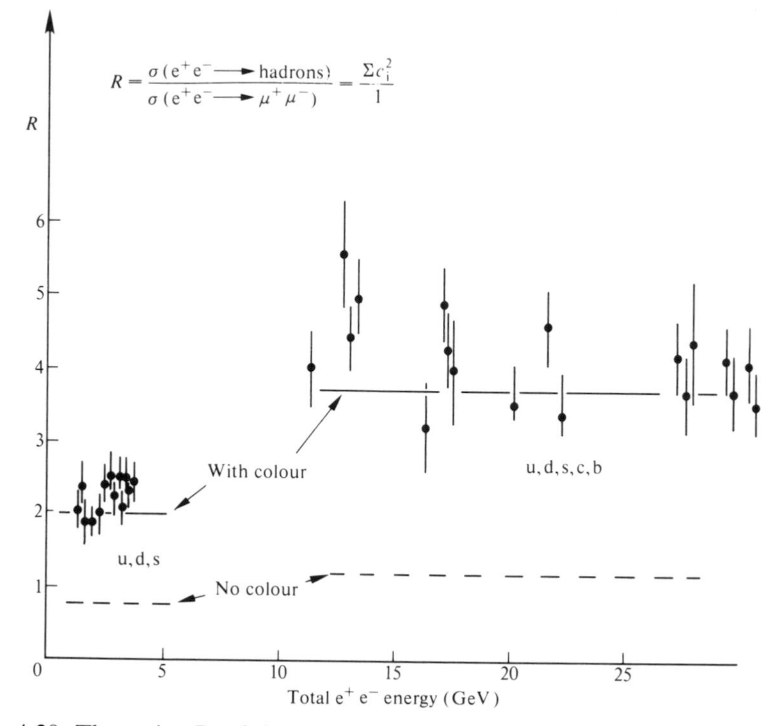

FIG. 4.20. The ratio, R, of the cross-sections for hadron and muon production in e^+e^- annihilation as a function of the total energy. There is an upward step of the expected magnitude above the thresholds for production of the c and b quarks and the data clearly favour the colour hypothesis.

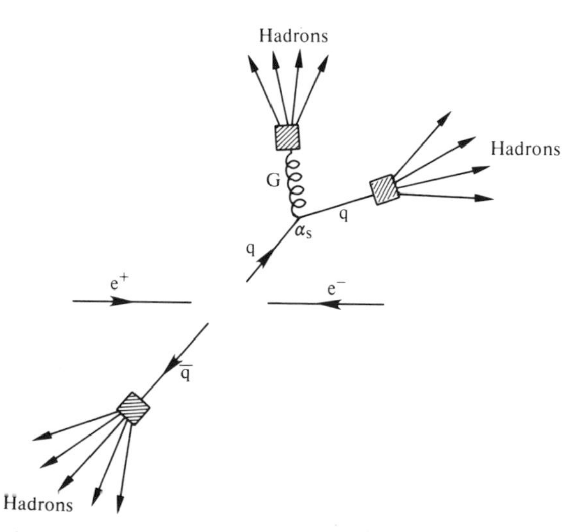

FIG. 4.21. Quark–antiquark creation followed by the 'radiation' of a gluon by one of the quarks should lead to a configuration with three jets of hadrons.

indeed to be seen, and an example is given in Fig. 4.22. Of course, although such observations are consistent with our expectations they do not prove the existence of gluons; the events might also be due to something else.

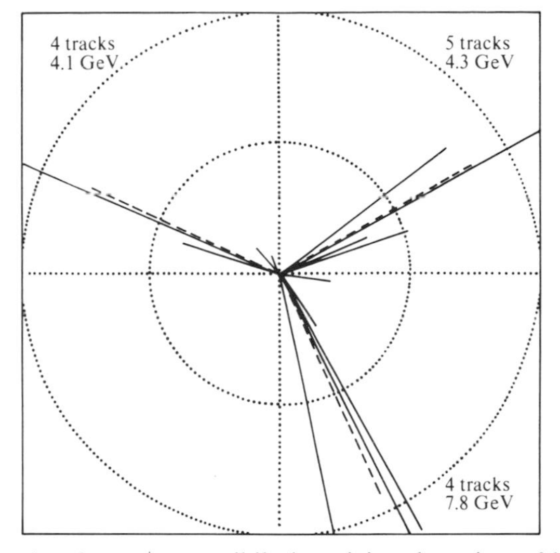

FIG. 4.22. Example of an e^+e^- annihilation giving three jets of hadrons. In this diagram the momentum of each particle detected in the TASSO experiment at PETRA is indicated by the length and direction of the lines.

Finally, the explanation of the 'missing mass' in Fig. 4.12 is straightforward in QCD theory. Part of the proton mass or energy has to be assigned to gluons, which only have strong colour interactions with other gluons or quarks, and do not interact with the electric or weak charges of the electrons and neutrinos respectively. Fig. 4.23 shows recent data on the integral under Fig. 4.12, plotted as a function of momentum transfer. The trend in the data is again consistent with the prediction but still does not at this stage prove the theory is correct. At best therefore, we can only say at present that QCD is a very good candidate theory of quark interactions: it comes through every experimental test, made in the region where the theory can be expected to apply, with flying colours.

Inside the proton

If this theoretical model is assumed correct, then our picture of the inside of the proton is indeed a bizarre one. What we see inside the proton depends entirely on the scale on which one observed it. I have tried to indicate this by three drawings of a proton. Let me warn you that although the whole human race loves to make pictures—because visual impression is an invaluable aid to comprehension—one must beware of taking these too literally. Pictures of macroscopic objects, the Moon or Jupiter for example, can be believed, but visual aids become suspect on a sub-nuclear scale, where quantum

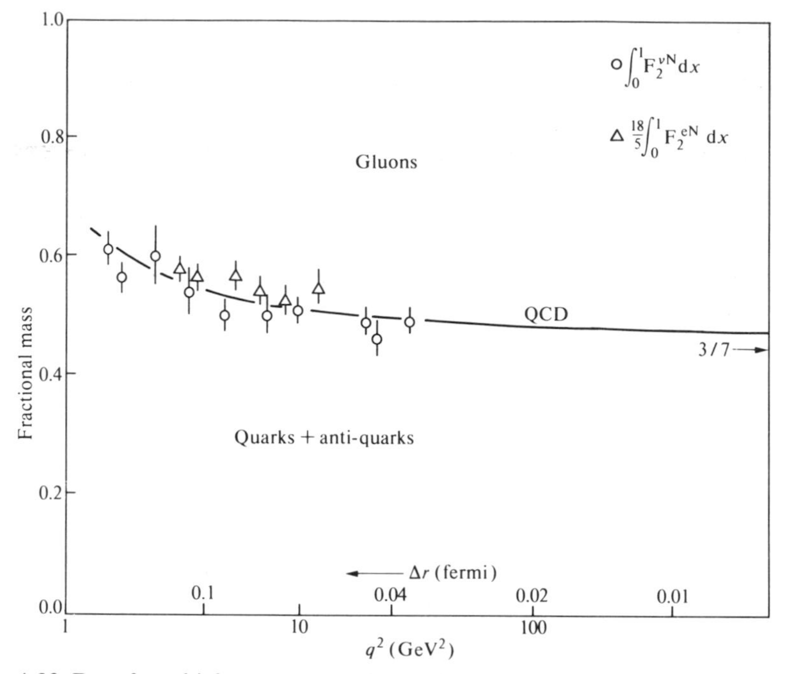

FIG. 4.23. Data from high-energy neutrino and electron scattering experiments show how the mass of the proton is shared between gluons and quarks. The dependence on q^2 agrees with the prediction of QCD.

effects really defy a pictorial description. The physics is contained in mathematical equations and in numbers.

Having said this, let us make some drawings nevertheless. The first (Fig. 4.24) shows a proton observed with radiation of a wavelength (or inverse momentum transfer) of a few times 10^{-13} cm, comparable to the proton size. At this level of resolution, the proton is just a pointlike, featureless blob. It might contain all sorts of things but the observational results, the only things you can believe, show no structure at all. This is what you would get in, say, the scattering of 50 MeV neutrinos (wavelength 4×10^{-13} cm) through large angles by a proton or neutron.

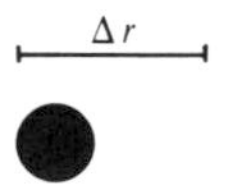

FIG. 4.24. The proton 'seen' with radiation of a wavelength about 4 fm (4×10^{-13} cm).

Next, we boost the beam energy to a few GeV. Now the resolution of our probe has improved, it can detect structure down to about one-tenth the proton size. We 'see' quarks directly, they scattered the neutrino, but we infer the existence of gluons because of missing mass. Occasionally, a gluon can transform to a quark–antiquark pair, so the picture is a little more complicated (Fig. 4.25). There is actual experimental evidence that a proton, at this resolving power, does contain antiparticles. We observe from Fig. 4.9 that the ratio of antineutrino to neutrino cross-sections, is about 0.42. *If* the proton contained quarks only, the weak interaction theory tells us that the ratio should be 1/3, because the cross-section for a neutrino to scatter off a quark is three times that for an antineutrino. Conversely, the cross-section for an antineutrino to scatter off an antiquark is three times that for a neutrino. So the antineutrino/neutrino cross-section ratio has to be between 1/3 and 3, and the actual number tells us the relative amounts of antiquarks and quarks. We can even plot their x distributions and these are shown in Fig. 4.26, at a value of momentum transfer, or spatial resolution, where antiquarks account for about 5 per cent of the total mass of the proton.

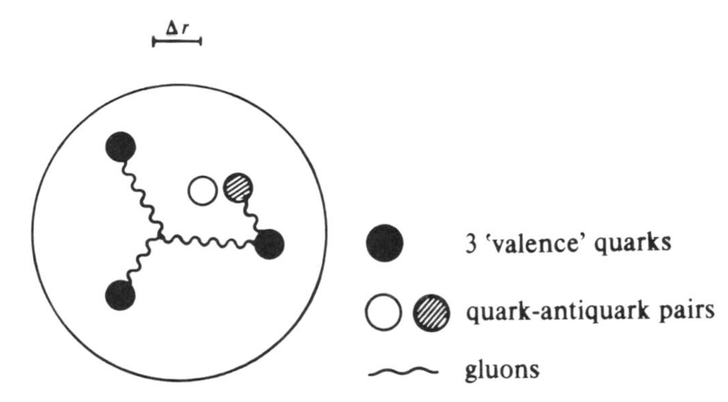

FIG. 4.25. The proton 'seen' with radiation able to resolve structure at 0.1 fm (10^{-14} cm).

Finally, we could go to a resolution some ten times greater, probing distances down to about 1/100 of the proton size (this is about the limit attainable with the present accelerators). In this case (Fig. 4.27) we find the inside of the proton seething with activity. What in Fig. 4.25 were a few quarks and gluons are now 'seen' to consist of yet more quarks and antiquarks and yet more gluons; indeed the original three quarks have been practically lost and contribute only a small fraction of the proton mass.

In a few sentences, a proton consists of quarks and gluons. And what do these consist of? Still more quarks and gluons—ad infinitum! The picture is very similar to that of the electron, consisting of an electron, virtual photons, virtual electron–positron pairs, as in Fig. 4.13. Is this bizarre picture correct, or is there a deeper underlying substructure? We don't know. We have no evidence at all for anything more fundamental, at present.

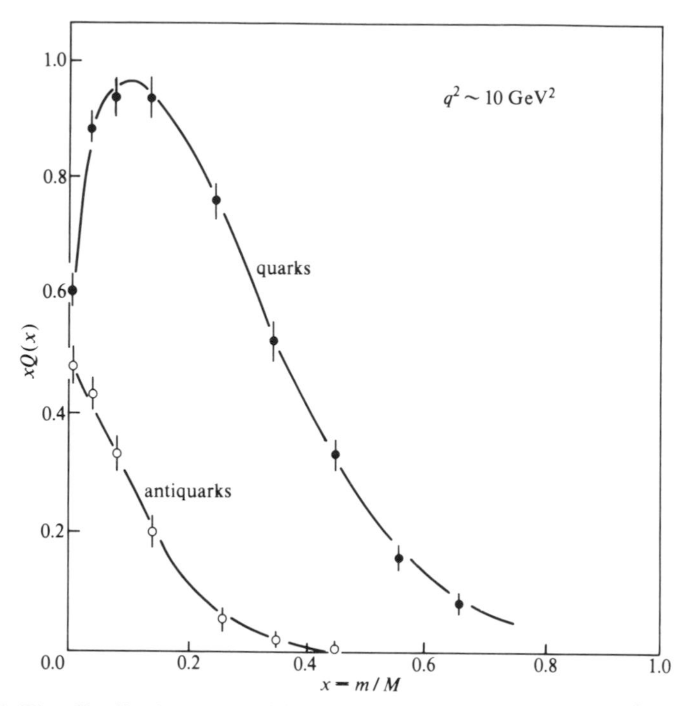

FIG. 4.26. The distributions in $xQ(x)$ for quarks and antiquarks at a $q^2 \sim 10$ GeV^2. The antiquarks account for about 5 per cent of the total mass of the proton at this q^2.

Protons are not forever (?)

There is one last picture of the proton which I must show you. It is quite different and perhaps more interesting, and it might be even less valid than the others. Fig. 4.28 shows a blank—a proton which has disappeared! Nothing, we are told, lives for ever, and we expect the proton to decay eventually. The predicted lifetime is very long, about 10^{32} years, so that in 1000 tons of protons, only 3 will decay in a year. The formidable task of searching for evidence of proton decay is currently being undertaken in several experiments deep underground (to reduce cosmic ray background). If the lifetime really is 10^{32} years, they should be able to find evidence for the decay process in the course of a few years.

Why should the proton decay and why is such a process important? After all, our universe has only been in existence for 10^{10} years, human (or any other) life has existed and will survive for only a tiny fraction of that time, and 10^{32} years is an eternity on such a scale. The reason is, we believe, connected with the whole nature of the way the Universe has developed. Excluding gravity, which has no significant affect on particles, at the energies available at present accelerators we distinguish three types of interaction, strong, electromagnetic, and weak, with coupling strengths of the order of 1, 10^{-2}, and 10^{-5} respectively. It is firmly believed that electromag-

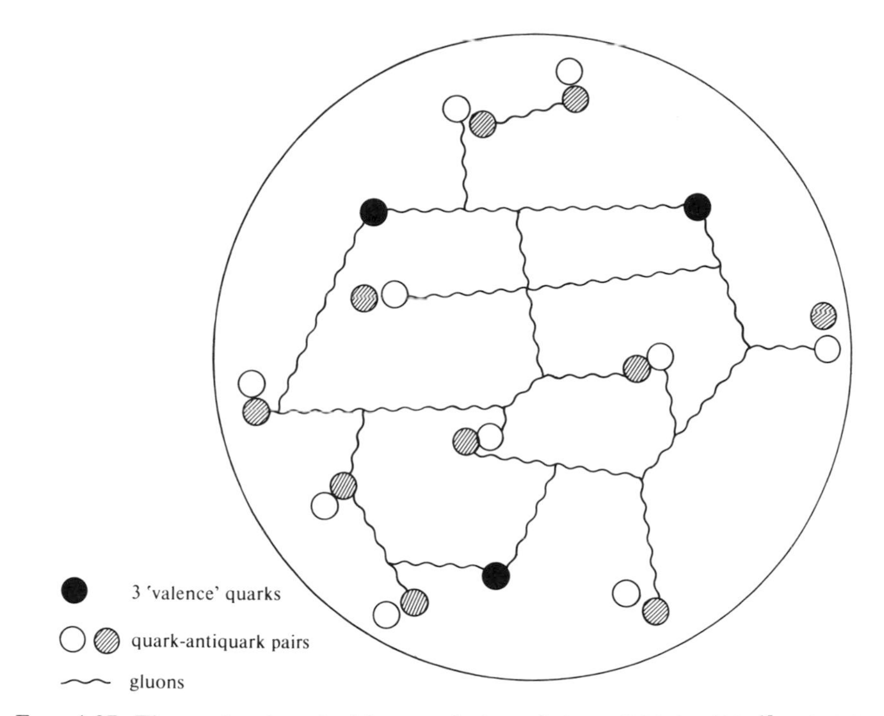

FIG. 4.27. The proton 'seen' with a resolution of about 0.01 fm (10^{-15} cm). The three 'valence' quarks are almost indistinguishable in a riot of gluons and quark–antiquark pairs.

FIG. 4.28. A proton that has disappeared.

netic and weak interactions are unified at sufficiently large energies. (Just recall from Fig. 4.9 that the weak cross-sections are increasing fast at modest energies.) They are considered as different aspects of a so-called electroweak interaction mediated by the photon and 'intermediate vector bosons' of mass almost one hundred times greater than the proton. Our discussion of proton structure in terms of QCD has, at the same time, shown strong interactions, those between the quarks, getting weaker at high energy, and clearly at some enormously high energy they could become comparable with electromagnetic and weak interactions. Miraculously, if one uses the data available at low energy and does this fantastic extrapolation, the three types of interaction seem to have merged at an energy $\sim 10^{14}$ GeV. In other words, it is possible that a higher symmetry exists, where new very massive bosons of mass $\sim 10^{14}$ GeV mediate a grand unified interaction.

In this scheme, strong, electromagnetic, and weak forces would, above 10^{14} GeV, become different aspects of one universal interaction and lose their separate identities. Quarks, previously distinguished by their strong interactions, would have the same couplings to the universal force as the leptons, before only experiencing the electro-weak force. One enters a strange new world of lepto-quarks, the massive bosons mediating the forces of these grand unified theories, that can transform quarks to leptons and quarks to antiquarks. Thus a proton, consisting of three quarks, can transform to a lepton and a meson (quark–antiquark pair), so that it is predicted to decay, for example, in the mode $p \rightarrow e^+ + \pi^0$. We are of course, nowhere near this energy in terrestrial conditions; the proton can only decay via the agency of a virtual lepto-quark, but because of the huge masses of these bosons the decay is very strongly suppressed. The theory also has other appealing features. It explains naturally the fractional quark charges (basically, because only when summed over all three colours should one obtain an integral charge), and the exact equality of electron and proton charges. But the only directly testable consequence of the model seems to be proton decay.

The Universe now is expanding and is generally assumed to have started as a 'Big Bang'. An enormous amount of energy was concentrated in a tiny volume which exploded. Matter was created out of energy. From symmetry arguments alone, we would expect equal amounts of protons and antiprotons. But the known Universe does not look like that at all. Our galactic messengers, the primary cosmic rays, are all nuclei, not antinuclei. Nor is there evidence for large amounts of antimatter in other galaxies: if there were, annihilation would take place with matter and enormously strong sources of gamma radiation should be seen—but they are not there. Something must have happened to produce a huge asymmetry between matter and antimatter.

The massive lepto-quarks of the grand unified theories may have played a vital role in these early events, a topic discussed in chapter 6 by John Ellis. It also turns out that another necessary ingredient is at least three pairs of quark flavours and three pairs of lepton flavours. The profligacy of Nature in providing those 'xerox copies' in Table 4.1, which seem so useless now, may therefore have an explanation. Only if it turns out that there exist more than six flavours of quarks and of leptons can we truly claim that we live in the best of all possible worlds. Otherwise, we are living in the only possible one!

So we return to the muon. We don't really know why it is there, and we can only speculate that such 'extra' particles had to be there if the Universe was to evolve in the way that it did. Fifteen thousand million years ago, the muon, like the other 'extra' leptons and quarks, must have been a star performer; now it is just a relic. But it is an interesting thought that, had the muon not been there then, we probably wouldn't be here now! A Universe evolving with only u and d quarks and e and ν_e leptons would, according to these ideas, have shown complete symmetry between matter and anti-matter. The number of protons and neutrons would have been only a tiny fraction of what is present now, and probably life as we know it could not have developed.

In conclusion, our quest into the internal structure of the proton, which started almost 50 years ago, has resulted in a somewhat bizarre picture filled with objects such as quarks and gluons which by their very nature can never be seen as free particles. Even more importantly, it has led us to try to understand the relations between the fundamental constituents of matter and their interactions, and furnished us with a host of new particles which seem to have no purpose in our present world. Each question raised has led to a solution which raises even more questions, and has widened the scope of our enquiry from the narrow confines of the proton and neutron to deep cosmological questions on the nature and evolution of our Universe.

It seems not inappropriate to close this lecture with a quotation from Voltaire, who had a lot to say about most things. In this passage I like to think that he had in mind physicists trying to describe the inside of the proton:

> Les Philosophes qui font des systèmes sur la secrète construction de l'univers, sont comme nos voyageurs qui vont à Constantinople, et qui parlent du Sérail: Ils n'en ont vu que les dehors, et ils prétendent savoir ce que fait le Sultan avec ses Favorites'
>
> Voltaire: *Pensées Philosophiques* (1766)

5

Unification of the Forces

ABDUS SALAM

Throughout the long history of mankind's search for an understanding of the physical world we have always believed the ultimate solution will be elegant in its economy of basic concepts. There are two aspects of this search: one is the search for the elementary entities of which all *matter* is made; the second is the search for ideas which seek to unify our understanding of the *forces* which act between these elementary constituents. In this chapter we shall mainly be concerned with this second aspect, coming back at the end to a unification which seeks to comprehend both matter and the forces within one single conceptual framework.

History of the unification ideas

As a starting point it is appropriate, and topical, to mention Al-Biruni, a great physicist who flourished about the year 1000 in Afghanistan. He was one of the first to speak about a universality of the laws of physics, a subject which 600 years later was to be placed on an experimental footing by Galileo. Galileo looked through his telescope and saw mountains on the Moon, mountains that cast shadows over the sunlit surface; the laws of shadow-making by the Sun's light were the same on the Moon as on Earth, so this was the first experimental demonstration of the universality of physics throughout the Universe. The next great physicist whom one remembers in this context is Newton, born in the year Galileo died. He unified *gravity*: celestial gravity, the force which keeps the earth in its orbit around the sun, and terrestrial gravity, the force responsible for falling apples. This was the first great unification in physics.

Another 150 years after Newton, Faraday and Ampere achieved the next fundamental unification by linking magnetism with moving electric charges. So the two forces of electricity and magnetism became one, the force of *electromagnetism*. Fifty years later, Maxwell developed a complete mathematical framework for the description of electromagnetism and showed that accelerating charges emit a radiation: electromagnetic radiation, which he identified with light. The laboratory confirmation of the existence of electromagnetic radiation came 10 years after his death. Maxwell's work was a landmark in the development of a unified description of physical phenomena and we are now familiar with the complete spectrum of electromagnetic radiation: radio, radar, radiant heat, visible light, ultra-violet, X-rays, and γ-rays.

This century we come to Einstein (born, the year Maxwell died, 1879), whose genius unified the concepts of space and time. Even more important than this unification, Einstein saw that, treating space–time as a dynamical system, its curvature could be associated with the gravitational force postulated by Newton. This led to an understanding of cosmology as an expression of space–time dynamics, the idea of the Big Bang origin of an expanding Universe, and the discovery of the 3 K radiation, a relic of the Big Bang.

In the mid-1920s we had these two forces, gravity and electromagnetism, describing almost all that we knew about Nature. And, quite rightly, Einstein then entertained the dream that having comprehended all the phenomena of Nature in terms of these two forces, these two should be capable of unification as aspects of one single force. He devoted the last 35 years of his life to this epic task and since he had shown that gravitation could be understood as a geometrical property of space and time, in terms of its curvature, he sought another geometrical entity which might correspond to electrical charge. His path to unification of these two forces thus lay through the geometry of space–time. We shall return to this dream later.

Nuclear forces

What Einstein neglected, perhaps because they were not so clearly understood when he was a young man, were the nuclear forces. Let us consider for a moment the elementary constituents of matter, the 'wheels within wheels within wheels' as Feynman has called them. In 1935 the innermost 'wheels' were thought to be just two pairs of particles: proton and neutron; neutrino and electron. While the electrons and neutrinos (members of the light, *lepton*, family of particles) are still presumed to be elementary, we now have evidence that the protons and neutrons are built of still more elementary entities, the *quarks*, but for the present we shall ignore this sub-structure.

Two types of nuclear force operate between these particles: the so-called *weak* nuclear force, which will be our main topic, and the *strong* nuclear force. The weak force is responsible for β-radioactivity, the phenomenon of the decay of the neutron into a proton together with an electron and an anti-neutrino:

$$\mathrm{n} \rightarrow \mathrm{p} + \mathrm{e}^- + \bar{\nu}_\mathrm{e}.$$

The strong nuclear force on the other hand is responsible for the binding of neutrons and protons within atomic nuclei, and the release of energy in nuclear fission and fusion. Both these forces are of very short range compared to electromagnetism and gravity. They only manifest themselves when these particles are close to each other: around 10^{-13} cm in the case of the strong force and less than 10^{-15} cm for the weak force.

Our present picture of how these forces operate is through the exchange of certain quanta. In the case of electromagnetism the quanta are the photons, the quanta of light. This exchange of quanta can be illustrated using diagrams named after Feynman, who introduced them.

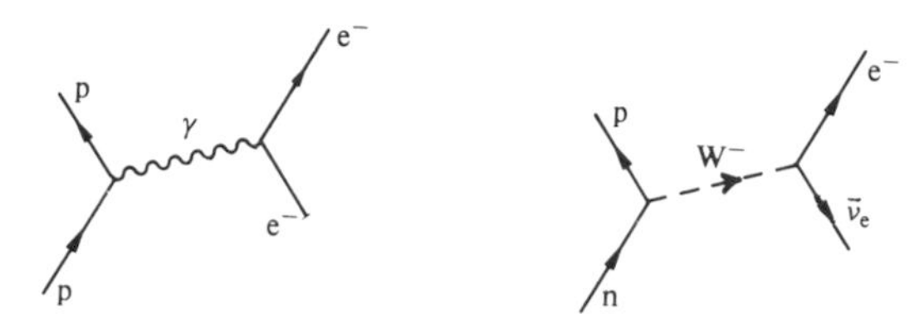

FIG. 5.1. Two interactions mediated by exchanged particles. The electromagnetic example shows the binding of an electron and a proton by photon exchange to form a hydrogen atom. The weak nuclear interaction is the β-decay of a neutron by W^- exchange.

In the case of the weak force we believe from indirect evidence that this force is mediated by quanta called W^+ and W^- (Fig. 5.1). We have not yet been able to produce these quanta in laboratory experiments, for reasons we shall speak about later; however one can infer from available evidence that these weak quanta must carry an electric charge (+1 or −1, in units of the electron charge); this is quite unlike the photon, which is electrically neutral. They should also be heavy in order to explain the manifest short range of the weak force, whereas the photon is massless.

One term I shall use frequently is *gauge force*. The mathematical definition of gauge forces is too technical for us to go into, but for our purposes it is sufficient to state that these are forces respecting a gauge symmetry and whose strength is proportional to a *charge*. For example the electromagnetic force is a gauge force and its strength is proportional to the product of the electric charges carried by the interacting particles, say the electron and the proton in a hydrogen atom. Likewise the strength of the gravitational force is proportional to the product of the two interacting masses; the mass is the charge for gravitation.

Just to show you how the ideas about fundamental forces change, I distinctly remember the lesson on the fundamental forces given to us by my first teacher of science in 1935 when I was a student at my birth place of Jhang in Pakistan. He told us about gravity and Newton's ideas; magnets were obtainable even in Jhang so he talked of magnetism as a fundamental force, then he said '. . . and now there is another force called electricity but that is found only in the capital city of Lahore, 150 miles to our east'; nuclear forces 'existed in Europe only'; but I still remember him telling us about the capillary force as one of the fundamental forces of nature. I had always felt surprised why he put so much emphasis on this force until I recalled that the physics he taught was a mixture of contemporary physics and the physics of Avicenna who wrote about the year 1000. Avicenna was both a physicist and a physician, so that any force which made blood rise into small vessels was clearly an exceedingly important force as far as he was concerned.

The unification of the weak force with electromagnetism

The story I am about to tell started, for me, in 1956. The problem that then concerned particle physicists was the nature of the weak force and, in particular, the question of the spin of the mediating quanta, the W^+ and W^- particles. In a rather round-about way our ideas led to the conclusion that the $W^{\pm}$ were objects carrying intrinsic angular momentum (spin) which in magnitude was the same as the spin of the photon: 1 in units of Planck's constant (they are often called intermediate vector bosons). This was the first significant pointer to some sort of connection between the electromagnetic and the weak forces.

These ideas came about in the following way. Dalitz had drawn attention to a perplexing situation in particle physics (called at the time the tau–theta puzzle). The details of this puzzle need not concern us, but it was taken up by Lee and Yang in the United States, who showed the riddle could be solved if the weak interaction discriminates between particles spinning in the left-handed versus the right-handed sense. That is to say, the weak force violates the symmetry between the left and the right or, equivalently, discriminates between an object and its mirror image. Mirror symmetry was known to be obeyed by all other forces and until then had been regarded as an axiomatic symmetry of Nature. Lee and Yang conjectured that electrons with a left-handed spin experience a weak force different from that of right-handed electrons. This conjecture of Lee and Yang would then, by a chain of arguments to be outlined below, lead to the proposition that the Ws have the same spin as the photon.

The next step, taken by John Ward and myself, and by Shelley Glashow of Harvard independently, in late 1958, was to suggest that if there is a connection between the two forces then the weak force, like the electromagnetic, should be a gauge force. Now this leads, by a mathematical argument, to the conclusion that you cannot have just two mediating particles, W^+ and W^-; there must be a triplet of these particles, the third one being electrically neutral. Let us call the third member of the postulated triplet W^0 for the moment.

As the third mediator must be electrically neutral, could it be the photon itself? If so, you would have an immediate unification of the weak and electromagnetic forces. Alternatively, if W^0 is not the photon, could it be a new 'heavy photon', similar to the photon but having a large mass like W^+ and W^-? In this case its existence would connote a new, hitherto undetected form of weak interaction. Such an interaction can be illustrated by a Feynman diagram showing, for example, a neutrino, ν, scattering off a neutron. The new weak interaction, mediated by the W^0, would have the characteristic property that the same set of particles (ν+n) are present both before and after the interaction (Fig. 5.2(a)). This is quite unlike the case of the only then known weak interactions, where (ν+n) in the initial state end up as e^- + p in the final state (Fig. 5.2(b)).

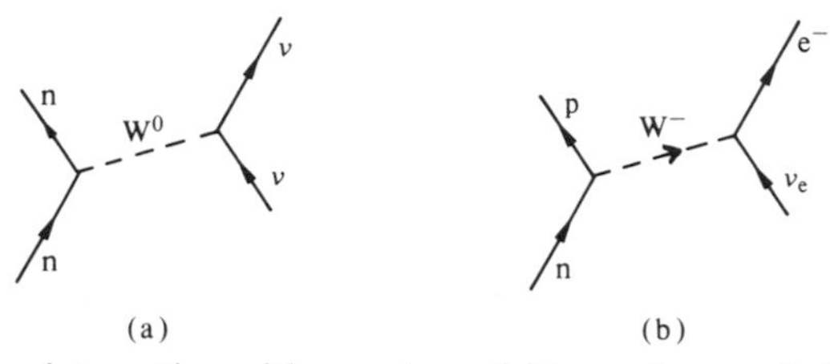

FIG. 5.2. Neutrino interaction with a neutron: (a) by exchange of a W^0 (the so-called neutral-current interaction); (b) exchange of a W^- (the first known form of the weak force, the charged-current interaction).

The first of the alternatives, in which the photon (γ) makes up a triplet together with two electrically charged mediators, W^+ and W^-, had already been considered in a somewhat different context in 1938 by Klein in Stockholm, but none of us knew of his work until 1979. For our generation the remark that W^+, W^-, and the photon might form a triplet was made by Schwinger in 1957, though he under-emphasized the gauge aspects of the theory, which were crucial to our way of thinking.

The second possibility of a triplet of three weakly interacting mediators, W^+, W^-, and the electrically neutral heavy mediator, nowadays called the Z^0, was first noted by Kemmer in 1937, but again without emphasizing the gauge aspects. These aspects (for this alternative) were emphasized by Bludman in 1958. Kemmer was my supervisor when I was a research student at Cambridge but I learnt of his paper only in 1972; thus we were ignorant of this also. The point of departure of our contribution was to show that the alternatives mentioned above are not really alternatives, nature does not do things by halves. The proposition of weak interactions violating left–right symmetry while electromagnetism obeys this symmetry, needed for its implementation not only W^+, W^-, and the photon, but also the Z^0, mediating a new form of weak interaction. This was the work of Glashow and later of Ward and myself, in the early nineteen-sixties. The weak and electromagnetic interactions could be unified, but at the same time a new type of weak force was predicted.

Let me go back to the question of the W spin and the clue which led us to consider gauge theories as candidates for a description of fundamental forces. These personal recollections formed part of the Nobel lecture which I was privileged to give on 8 December 1979 and for those of you who may become research scientists later on may contain some lessons on how not to be frightened by eminent authorities in your field! In September 1956, at the Seattle Conference, I heard Yang expound his and Lee's ideas on the possibility of the hitherto sacred principle of left–right symmetry being violated in the realm of the weak force, in response to Dalitz's puzzle. I remember, on an overnight flight back to London, reflecting on why nature should violate left–right symmetry in weak interactions. There came back to me a deeply perceptive question about the neutrino which Professor Rudolf Peierls had asked me when he was examinining me for a Ph.D. a few years before. Peierls' question was: 'The photon mass is zero because of Maxwell's principle of a gauge symmetry for electromagnetism. Tell me, why is the neutrino mass zero?'. I had then felt somewhat uncomfortable at Peierls asking me in a Ph.D. viva a question to which he himself said he did not know the answer! But during that comfortless night, an answer came. The analogue for the neutrino of the gauge symmetry for the photon existed. It had to do with the possible masslessness of the neutrino and an exact symmetry (which one technically calls chiral symmetry) which could exist for massless particles. However the existence of such a symmetry inevitably would mean that left–right symmetry must be violated. In my formulation there must be no left–right symmetry exhibited by the interactions where the neutrino was a participant.

I got off the plane the next morning naturally very elated. I rushed to the Cavendish, worked out a quantity relevant to muon decay, the Michel parameter, and a few other consequences of these ideas, dashed out again, got on a train to Birmingham where Peierls then lived. I presented my ideas to Peierls. He had asked the original question, could he approve the answer? Peierls' reply was kind but firm. He said 'I do not beleve that left–right symmetry can be violated in weak nuclear forces at all'. Thus rebuffed in Birmingham, like Zuleika Dobson, I wondered to whom I could go next and the obvious person was Pauli, the father of the neutrino, who at that time served as an oracle on the subject. I went to CERN at Geneva and sent Pauli my paper in the hands of Professor Villa of MIT who was to visit him at his home in Zurich. Soon the message came back. 'Give my regards to my friend Salam and tell him to think of something better'.

This was discouraging, but I was compensated by Pauli's excessive kindness a few months later when Mrs Wu, Lederman, and Telegdi performed their experiments and showed that Peierls and Pauli were wrong. Left–right symmetry was indeed violated in the weak interactions; neutrinos existed only in the left-handed state, just as I had predicted. I received Pauli's somewhat apologetic letter on 24 January 1957. Thinking that by now Pauli's spirit should be suitably crushed, I sent him two short notes I had written in the meantime. These contained suggestions to extend chiral symmetry to electrons and muons, assuming that their masses were consequences of what has come to be known as dynamically spontaneous symmetry breaking. With chiral symmetry one could show that the mediators of the weak force, the $W^{\pm}$s, must carry spin 1.

Reviving thus the notion of charged intermediaries ($W^{\pm}$) of spin 1 one could then postulate for these a type of gauge invariance which I then called the neutrino gauge. Pauli's reaction was swift and terrible. He wrote on 30 January 1957, then on 18 February and later on 11, 12, and 13 March (you see the telephone habit had not been adopted by then) 'I am reading (along the shores of Lake Zurich in bright sunshine) quietly your paper. I am very much startled by the title "Universal Fermi Interaction". For quite a while I have the rule that if a theoretician says universal it just means pure non-sense. This holds particularly in connection with the Fermi interaction and now you too Brutus, my son, come with this word.' Earlier, on 30 January, he had written, 'There is a similarity between this type of gauge invariance and that which was published by Yang and Mills', and he gave me the full reference of Yang and Mills' paper. I quote again from his letter, 'However, there are dark points in your paper regarding the vector field which you introduce, W^+ and W^-. If the rest mass is infinite or very large, how can this be compatible with the gauge transformation which you have postulated?' And he concludes his letter with the remark, 'Every reader will realize that you deliberately conceal here something and will ask you the same questions.' Although Pauli signed himself with 'friendly regards', he had forgotten his earlier penitence. He was clearly and rightly on the warpath.

Now the point which Pauli was making was absolutely important and crucial for the development of the theory. The point was this; here we are talking blithely of W^+, W^-, Z^0, and the photon as being aspects of the same fundamental force. If a perfect symmetry exists between these particles, then they would be like four components of one single entity and could be transformable one into the other under suitable conditions. However, we knew that electromagnetism was a long-range force, with an effect falling off as the inverse square of the distance, while the weak force was a very short-range force, falling off exponentially. How could we talk of a symmetry between these two totally different sorts of force: one long-rang and the other short-range?

One more idea was needed before we could credibly talk about a symmetry between these entities and it took the next seven years before the idea was finally realized. Chris Llewellyn-Smith has discussed this idea in chapter 3 but I shall outline it also. We needed something which could break the symmetry between the W^+, W^-, Z^0, and the photon, spontaneously, as it is nowadays called. This is a concept which had already been invented by physicists studying the solid state, and they had given it the much better name of 'order' rather than 'broken symmetry'; we needed to introduce the phenomenon of order in particle physics.

To describe how order and symmetry differ from each other, think of a table, a circular table, at which a number of people are sitting down for their meal. There are plates in front of them and there are napkins. Now suppose this is a country where the table conventions have not been fixed so that you don't know which napkin is yours and which is your neighbour's. This is the

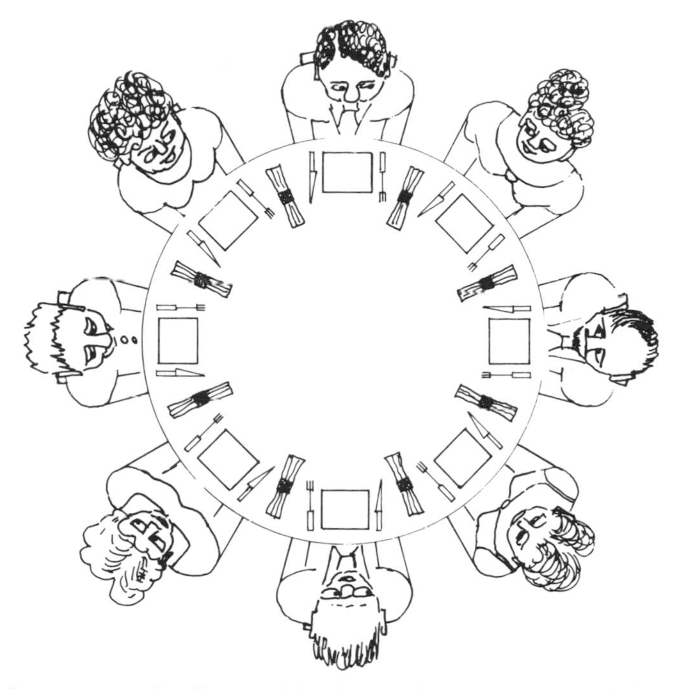

FIG. 5.3. Symmetry at the dinner table; choice of a napkin will break the symmetry and establish 'order'.

state of symmetry. You may have seen the situation, when neighbours are surreptitiously eyeing each other. Then somebody decides: this is my napkin. It does not matter if the napkin is to the left or the right. But someone makes the decision. Instantly the whole table gets 'ordered'; this is the phenomenon of *symmetry*, changing instantly (with the velocity of light) into *order*, precisely what we needed in particle physics. Order is broken symmetry: you have chosen to call the left-handed napkin yours, thus the symmetry has been broken but in an orderly fashion.

But to establish the state of order we must pay a price. We must start with a special sort of force-field which compels an asymmetrical choice. As an illustration, we can think of a situation in which the energy of the system under study rises if the system is displaced from its equilibrium position symmetrically in any direction. This is the 'begging-bowl' shape for what is called the 'potential function' (Fig. 5.4(a)). There is, however, the possibility of another type of potential we can call 'dimpled-bowl'. This has the necessary property of being a symmetrical potential function but is one which 'forces' an ordered solution with less symmetry than the begging-bowl type.

To conclude, there had to be, in addition to the gauge forces, a new type of force which would give rise to the dimpled-bowl type of potential. Such a force required the introduction of one more particle, the Higgs particle,

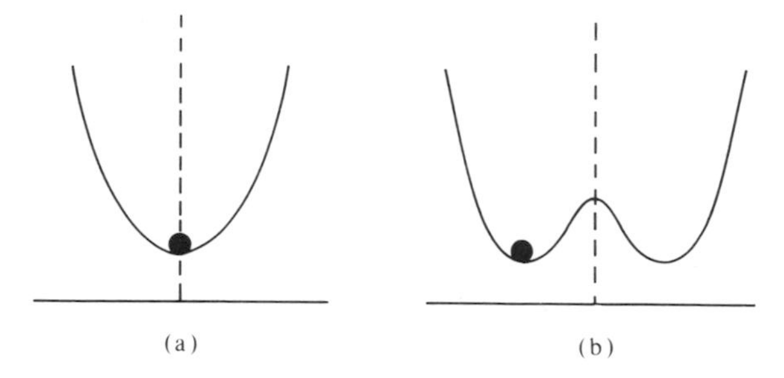

FIG. 5.4. Symmetric potential functions for which the stable solutions are (a) symmetric, (b) asymmetric.

named after the physicist who contributed to the invention of the necessary mechanism.

In 1967 Weinberg and I used these concepts of ordered symmetry breaking to build up a theory where the W^+, W^-, and Z^0 were heavy, so giving a short range to the weak interaction, while the photon had zero rest mass and the electric force was thus long-range. In our theory the ratio of the masses of the Z^0 and $W^\pm$ particles were fixed in terms of the parameters of the theory.

To summarize then, during 1956–57 experiments on left–right (or mirror) symmetry violation led us to suspect that the W^+, W^-, and the photon had the same spin. We conjectured that matters did not just rest with the identity of spins of the two types of particles. Perhaps there was a deeper identity, manifested by the common gauge symmetry character of the two types of forces. But then came the question of whether there should be an additional neutral heavy photon, Z^0, as well as the massless photon. From the fact that electromagnetism respected left–right symmetry and weak interactions did not, the answer had to be yes, implying a new form of the weak interaction mediated by the Z^0. Next came the problem of showing why W^+, W^-, and Z^0 were heavy while the photon was massless; the answer was spontaneous symmetry breaking, or the phenomenon of order, using the Higgs mechanism. This introduced a new particle, the Higgs particle with spin zero, and also led us to the prediction of definite values for the masses of the Ws and the Z^0.

A number of other crucial technical developments connected with a most desirable property of a theory, calculability, took place in the early 1970s, associated with the name of an exceptionally gifted young physicist from Holland, t'Hooft. These I will not discuss. But one early difficulty with the theory of the Z^0 interactions led to the prediction of a new quark. At that time only three quarks were known: u (up), d (down), and s (strange). A satisfactory theory of the Z^0 required that quarks came in doublets. The u and d quarks were fine but the strange quark was an orphan; there was need of a companion for it. This was postulated by Glashow and his colleagues, who named it the charmed quark (c) and we awaited the verdict of experiment.

In 1973 came the first verdict. Through the intermediary of the Z^0, neutrinos could interact with protons, or neutrons, without either the neutrino or the proton being changed, in contrast to the case for W^+ or W^- exchange (Fig. 5.2).

Likewise, the theory predicted that a neutrino and electron could scatter off each other. Such processes were among the predictions of the new theory but had not been seen. To give you some idea of the difficulties, the neutrino–electron scattering process has a cross-section of about $10^{-42}\,cm^2$, meaning that the probability of such an occurrence is about 15 orders of magnitude smaller than for processes induced by the strong nuclear force which form the substance of most experimentation in nuclear physics.

In 1973 at CERN the bubble chamber named Gargamelle detected both types of process in an epic experiment (Fig. 5.5). Its findings were soon confirmed by a number of other experiments and the clinching one was performed at SLAC, the Stanford linear accelerator, in 1978. In this experiment one observes electron scattering on protons; here there is not only a photon exchanged but also the Z^0 particle, as illustrated in the two Feynman diagrams of Fig. 5.6.

These two contributions to the scattering process combine in a specific and predictable way. In particular the violation of left–right symmetry inherent in the Z^0 contribution, although a very small effect, can be detected and leads to a slightly greater probability, about 1 part in 10 000, of a scattering taking place if the incident electrons are spinning left-handedly than if they are right-handed. This subtle effect has been carefully measured and is in excellent agreement with the theoretical predictions, as you can see from Fig. 5.7, showing the data obtained by Prescott, Taylor, and their collaborators.

The second verdict came shortly afterwards, in 1974, with the discovery by Richter and Ting and their colleagues of a new flavour of quark with all the properties expected for 'charm'.

At this point I would have liked to describe the experiments which have been done and pay a tribute to the ingenuity, persistence, and skills of the experimenters; the tremendous scale of present-day particle physics experiments is fantastic, especially when you compare them to the apparatus used by Hertz (Fig. 5.8) to demonstrate the existence of the electromagnetic radiation predicted by Maxwell. To test the unified theory of the weak and electromagnetic forces takes us to much higher energies, enormous particle accelerators and massive pieces of experimental apparatus like those shown in Figs. 4.6 and 4.7.

After the discovery of this indirect evidence for the existence of the Z^0, the next most important step is to produce and detect the Z^0 itself, and this prize is attracting intense effort on the part of accelerator builders and experimenters everywhere. At CERN a programme of work is under way

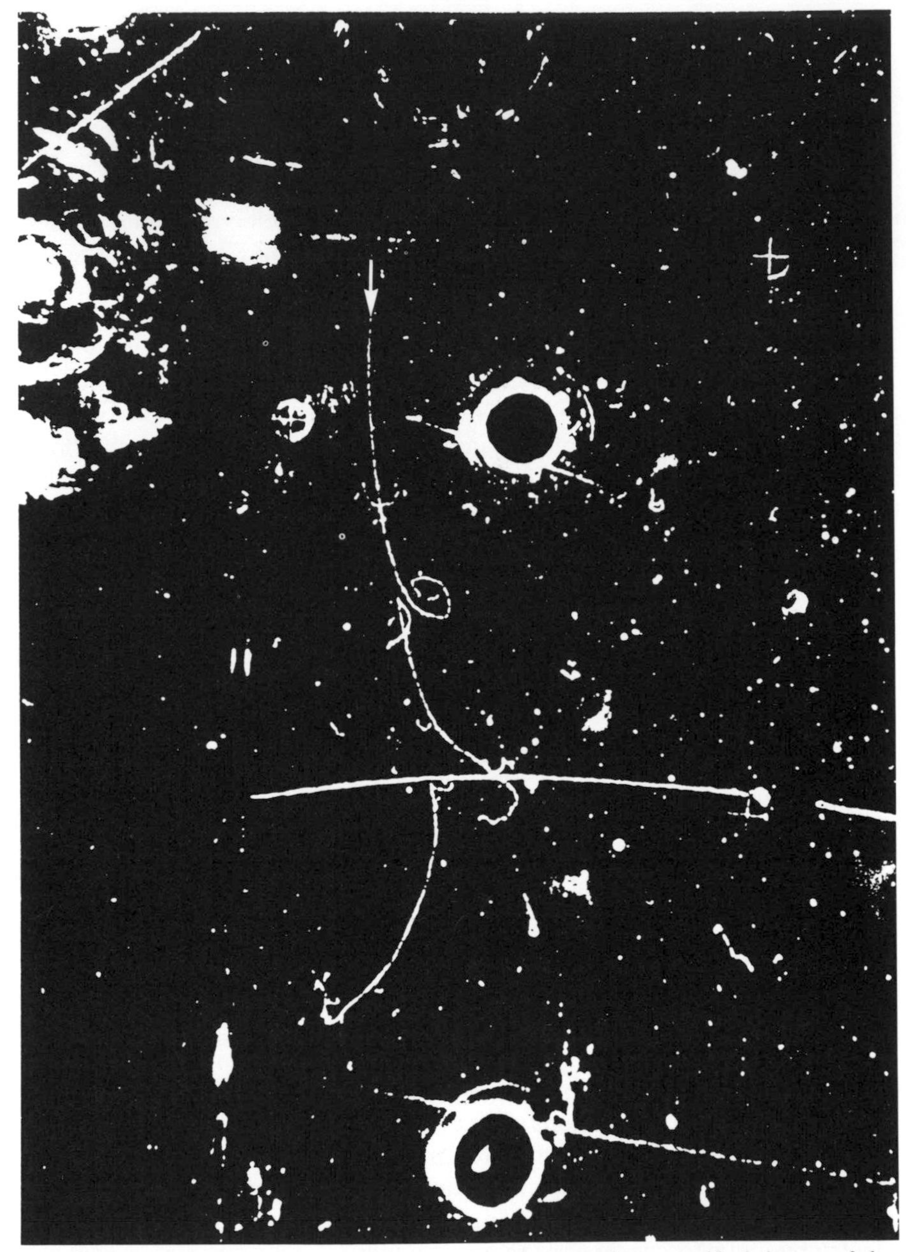

FIG. 5.5. A crucial experimental observation in the path to a unified theory of the weak and electromagnetic forces; in this picture taken when the Gargamelle bubble chamber at CERN was exposed to an intense beam of neutrinos the track of an electron recoiling from being struck by a neutrino is seen. This reaction:

$$\nu_\mu + e^- \rightarrow \nu_\mu + e^-$$

requires the existence of a weak force mediated by a neutral exchange particle, the Z^0.

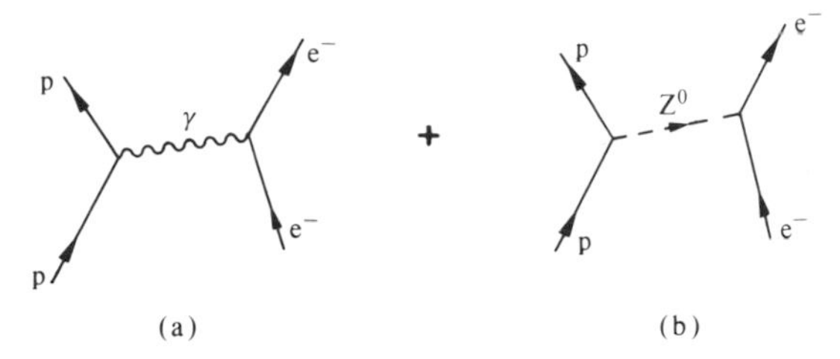

FIG. 5.6. Electron–proton scattering. Two diagrams contribute: (a) the electromagnetic force by photon exchange; (b) the neutral weak force by Z^0 exchange.

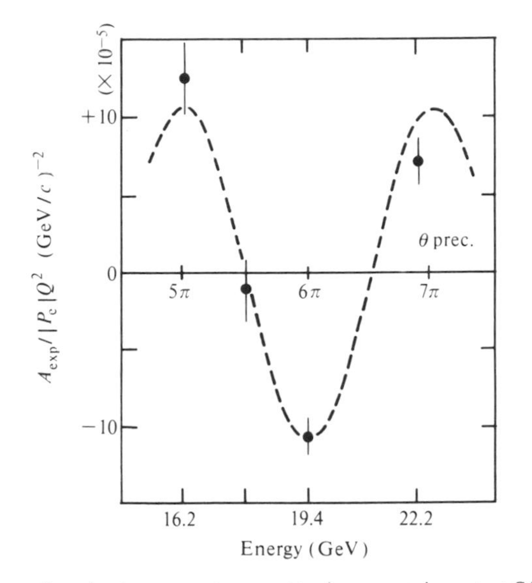

FIG. 5.7. Data from the electron–proton scattering experiment at SLAC compared with the prediction of the electro-weak theory.

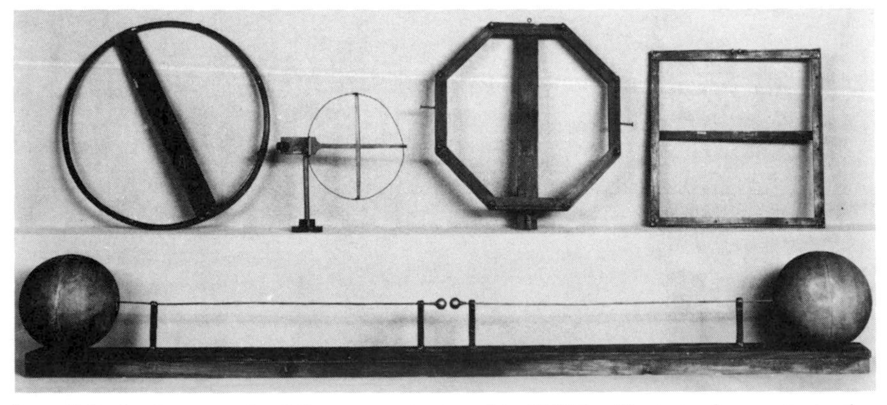

FIG. 5.8. A photograph of the apparatus used in 1888 by Hertz to demonstrate the existence of the electromagnetic radiation predicted by Maxwell. (Deutsches Museum, Munich.)

which, by the middle of 1981, will enable the large proton accelerator, the SPS, to store counter-rotating beams of protons and anti-protons at an energy of 270 GeV each (see chapter 7). When these collide, head-on, each annihilation will release 540 GeV, amply sufficient for the creation of Z^0 at its predicted mass of 90 GeV. Although more difficult to detect, the charged mediators W^+ and W^- with masses about 80 GeV should also be produced. The experiments this time are on an even grander scale; Fig. 5.9 shows one of the largest being prepared to hunt the Z^0 at the CERN machine.

A colliding beam machine for protons, of greater beam energy and intensity, is under construction in the USA, at Brookhaven, and Fermilab is to follow a similar line to CERN but at higher energy. Again in Europe, but on a longer time-scale, it is proposed to build the world's biggest electron–positron collider, called LEP; it would have a circumference of about 30 kilometres and would produce enormous numbers of Z^0 for a detailed study

FIG. 5.9. A photograph of the apparatus for experiment UA1 at CERN being prepared for the Z^0 search using the new proton–antiproton storage mode of the SPS. (CERN.)

of many aspects of physics. It would be capable of going to higher energies for further crucial tests of the unified theory of gauge forces (John Adams describes this enterprise in chapter 7). The two-mile long electron accelerator at SLAC, which performed the important photon–Z^0 interference experiment I have just described, may also be used to accelerate electrons and positrons which could be made to collide and create a useful number of Z^0s.

So the search for the Z^0 is generating great interest and excitement at the present time. The success of the theory so far has given us considerable confidence in the mass predictions. For example, the theory relates the masses of the $W^{\pm}$ and Z^0 to a quantity I shall call R, which is measurable in high-energy neutrino experiments; the theoretical prediction for R is 1 and the latest experimental value is 1.00 ± 0.02, a fantastic agreement when one considers the problems associated with the detection of neutrinos, with their very, very weak interactions. But it is still necessary to produce the Z^0, and $W^{\pm}$, in the laboratory, to study them and the detailed production process; as electromagnetic radiation was produced and studied by Hertz.

There is an apocryphal story about Einstein, that, when asked what he would have thought if experiment had not confirmed his prediction of the deflection of light from a distant star by the gravitational field of the Sun (observed at a Solar eclipse in May 1919), he replied: 'Madam, I would have thought the Lord had missed a most marvellous opportunity'. I believe, however, that the following quote from Einstein's Herbert Spencer lecture of 1933 at Oxford expresses his, my colleagues', and my own views more accurately: 'Pure logical thinking cannot yield us any knowledge of the empirical world; all knowledge of reality starts from experience and ends in it'. This is exactly how I feel about the Gargamelle–SLAC experience and how I now await the results of the Z^0 search.

Unification of the electro-weak force with the strong force: the electro-nuclear force

So far we have been considering unification of the electromagnetic force and the weak nuclear force; now let us turn to the possibilities of unifying these with the other forces of nature. We started with four fundamental forces and believe we have gone down to three; can we now make the next step, to go down to two by combining the electro-weak with the strong nuclear force, that is, formulate a theory of an electro-nuclear force? Here again the important idea is the concept of gauge symmetry, which relates the strength of the interaction to a charge. Now, with recent developments in the theory of strong interactions, there is a candidate for 'strong charges', known as 'colour charges'. The theory of the strong nuclear force embodying colour charges, carried by quarks and by the gluons (which, it is proposed, mediate the strong force), is a gauge theory known as Quantum Chromo-dynamics or QCD. The gluons, like the photon, $W^{\pm}$, and Z^0, have spin 1, and we believe QCD can, in principle, be combined with the electro-weak gauge theory to achieve a complete unification of electromagnetism and the weak and strong nuclear forces into an electro-nuclear force.

The crucial hallmark of this idea is the decay of the proton, discussed by Jogesh Pati and myself in 1973 and Howard Georgi and Shelley Glashow in 1975. The strong interaction operates between quarks while the weak force exists between leptons and quarks; if these forces are really the same then quarks and leptons should also be aspects of one single entity. It should then be possible for a component of the unifying force to transform quarks into leptons, and this means the proton (made of quarks) cannot be absolutely stable. The observation of its lack of stability, the decay of a proton into leptons, would be the test of the possibility of unification of the strong nuclear force with the electro-weak force, and many physicists are now looking for ways to detect proton decay.

But we already know this process must be very rare, if it can happen at all. The first to put a lower limit to the proton lifetime was Goldhaber with a 'back of envelope' calculation; he argued that if our life span is three score years and ten, or thereabouts, and if the protons (and neutrons) in our bodies are decaying all the time, the radioactivity must not be so great, over our life-span, as to cause serious degeneration of our bones! The maximum radioactive dose we could tolerate enabled him to put a lower limit of about 10^{16} years to the proton lifetime; if it were shorter the radioactivity generated by the decay of our own protons would cause significant deterioration to body tissues over a period of 70 years. Considering that the Universe has only existed for 10^{10} years, you can see that this is already an enormously long time!

The limit was raised by Reines and his collaborators, who experimented in a South African gold mine one and a half miles deep; they claimed a lower limit of 10^{30} years, unimaginably long. At such a lifetime, the Earth, which has existed something like 10^{9} years, would by now have lost about 10 tons of material by proton decay. Within the next few years we should have an answer to this most fundamental question; several experiments are being prepared in the USA and others in Europe; although these experiments are very difficult (and of large scale in both space and time!) the present experimental limit, 10^{30} years, is not tremendously far from what we now believe, theoretically, to be a possible figure. These efforts may soon tell us whether we can take the next step of electro-nuclear unification.

Einstein's dream; unifying gravity with the electro-nuclear force

Finally, we come back to Einstein's dream, now in the form of an attempt to unify the four forces to make one, by joining gravitation into the same complex of gauge forces as electromagnetism, the weak and the strong nuclear forces.

This is going to be the hardest step of all, and the intellectual triumph of

Einstein's theory of gravitation is itself a demonstration of the magnitude of the task of confronting us. The secret of Einstein's achievement, I believe one of the greatest mankind has known, at least in physics, is that he understood the fundamental meaning of *charge* for the gravitational force. The point I wish to make is that not until we understand, as deeply as Einstein did for gravity, the nature of the charges for the electromagnetic, weak, and strong forces, can we hope to succeed in this final unification.

Einstein found that the *gravitational charge* could be represented in terms of curvature in a four-dimensional manifold of space and time, and this gravitational charge was the same as *inertial mass*. This is Einstein's *principle of equivalance*, equating two very different sorts of thing, and it is fundamental, a keystone in his theory. It is also a remarkably precisely tested principle, as we shall now see. Let us take, as example, a hydrogen atom; this is made of a proton and an electron, with masses we shall denote by m_p and m_e. But the mass of the hydrogen atom is not $(m_p + m_e)$, it is a little less, because when the proton and electron come together, forming the atom, the total energy of the system is a little smaller than it was when they were separated. The difference is the 'binding energy', E; if you want to pull the atom apart and separate the proton and electron again this is the amount of energy you have to put back into the system. But, since mass and energy are equivalent ($E = mc^2$) this fall in total energy when the hydrogen atom is formed is manifest as a very small decrease in the mass of the whole system.

The point at issue is, does the gravitational charge for the hydrogen atom correspond to $(m_p + m_e)$ or does it (as determined experimentally by the magnitude of the gravitational attraction of two hydrogen atoms) exactly follow the inertial mass and equal $(m_p + m_e - E/c^2)$? The electromagnetic binding energy, E, corresponds to a mass decrease of (E/c^2), which is of the order of 10^{-8} of the hydrogen atom mass. That the principle of equivalence requires this small correction to be made also to the gravitational charge was shown by the experiments at Eötvös to 1 part in 10^9, later improved by Dicke, and then Braginsky and Panov, to 1 part in 10^{12}. This accuracy tests the equivalence principle so far as the binding energy attributable to electromagnetism is concerned.

But the gravitational force itself also contributes minutely to the binding of the hydrogen atom. Should we also subtract the effect of this from the gravitational charge? Einstein's principle of equivalence says yes, and this is crucial to his theory. Dicke and others have invented theories in which this last modification to the inertial mass is not followed by the gravitational charge. However a test seemed unattainable, since the gravitational binding energy is no more than 10^{-47} of the hydrogen atom mass and this is far beyond the ingenuity of any experimentalist to detect! But if one uses large enough test bodies, say the Earth and the Moon, the gravitational binding energy is as much as 10^{-10} of the total mass and this is just about measurable, provided someone is kind enough to go to the Moon and place a mirror there! In the three-body system of Sun, Earth, and Moon, as the Moon orbits the Earth any difference between inertial mass and gravitational charge will affect their relative motion in a way which can be detected by accurately measuring the variation in the distance between the Earth and the Moon. Such measurements were made in 1976, using a laser beam reflected by the mirror, by two groups of experimenters, one led by Shapiro and the other by Dicke himself. The distance between the Earth and the Moon was measured within an error of ± 30 cm and the result was another outright success for Einstein: inertial mass and gravitational charge are the same to about 1 part in 10^{11}, including gravitational binding energy.

This illustration of the success of Einstein's theory emphasizes the extreme quantitative precision of his comprehension of the gravitational charge as a manifestation of curvature of space–time and of its equivalence to inertial mass. It was Einstein's dream to unify the electromagnetic force, responsible, as he saw it, for the macroscopic properties of matter, with gravitation; to achieve this he devoted his last 35 years to the search for a meaning of the electric charge in terms of properties of space and time, just as he had found such a basis for gravitational charge.

We would like to take up this task, which defeated Einstein, and also do for the other charges (that is those for the weak nuclear force and the strong nuclear force) what he did for the gravitational charge. One attempt at a unification of gravity and electromagnetism was made in the early twenties by Kaluza and Klein who imagined a five-dimensional world and computed the curvature in five dimensions, just as Einstein had done in his four dimensions of ordinary space plus time. The amazing result was that the extra equations involving the components of the curvature in the fifth dimension were nothing else but Maxwell's electromagnetism. Thus the electric charge may possibly be connected with a fifth dimension and the curvature corresponding to it!

What do we make of this? There has recently been a revival of this suggestion of Kaluza and Klein by Cremmer and Scherk and others; perhaps the extra dimension really does exist. We are unconscious of it because it was 'curled up' to a very small radius, less than 10^{-33} cm, in the first 10^{-43} seconds of the world's birth in the Big Bang. Why did that happen? Again a choice of a particular 'order', or a spontaneous symmetry breaking due to a suitable potential, reduced the higher dimensional space–time to the four-dimensional one we are conscious of; the fifth dimension is effectively hidden from us, except indirectly through the manifestation of electric charge. This is one idea for uniting gravity and electromagnetism, but it is an empty idea until we can suggest some way which would test it.

A second idea is due to Wheeler and Schemberg, who posited that electric charge may not be connected with curvature in extra dimensions but is a manifestation of the topological structure of space–time, with 'worm-holes' in space–time, a sort of 'Gruyere-cheesiness'. The proposition is that if you go down to 10^{-33} cm, space–time is no longer continuous, there is a specific type of granularity. This idea has recently been developed by Hawking and his collaborators. Again, one has to find ways to test the idea.

But we must remember that our task is not just to find a representation of the electric charge. There are the other gauge forces we have been discussing and charges corresponding to these. Does this mean an extension to even higher numbers of dimensions? And if so, how many dimensions? To answer this question we must count how many different types of charges we seem to need.

The present list of elementary particles contains three 'families' of quarks and leptons:

$$\text{quarks:} \quad \begin{pmatrix} \text{u} \\ \text{d} \end{pmatrix}, \begin{pmatrix} \text{c} \\ \text{s} \end{pmatrix}, \begin{pmatrix} \text{t(?)} \\ \text{b} \end{pmatrix}$$

$$\text{leptons:} \quad \begin{pmatrix} \text{e} \\ \nu_e \end{pmatrix}, \begin{pmatrix} \mu \\ \nu_\mu \end{pmatrix}, \begin{pmatrix} \tau \\ \nu_\tau \end{pmatrix}.$$

Both quarks and leptons come in 'flavour-charge' pairs and in addition the quarks carry one of the 3 'colour-charges'. The total number of these entities is thus 24. (That is: quarks $3 \times 2 \times 3 = 18$, plus leptons $3 \times 2 = 6$, makes 24.) This is a large number and one is tempted to ask whether they could all be formed from still more elementary forms of matter, let us call them *pre-quarks* or *preons*, each carrying one of 8 fundamental charges: 3 for colour, 2 for flavour, and 3 for family distinction. There is no evidence for this (and it may be especially hard to imagine a massless neutrino as a composite particle), but the idea has a number of attractions. Indeed, it has been suggested that such preons may carry a magnetic charge in addition, that is, they are magnetic monopoles, and the very strong binding of opposite (North, South) pairs could form the 24 quarks and leptons listed above; the tight binding of this new sub-structure might perhaps be separable at energies greater than about 10^5 GeV. The presence of monopolies in the theory would also complete a symmetry in the Maxwell equations of electromagnetism and, as first pointed out by Dirac, explain the quantization of electric charge.

If we imagine that the total number of charges is eight, do we need to add more dimensions to our familiar space and time if we wish to extend the work of Kaluza and Klein, who were concerned with uniting just one charge (electric) with the gravitational one? It turns out that we need eleven dimensions to accommodate the eight charges.

Unifying forces and matter: super-gravity

But is this unification of forces enough? Would we not like to have a unified description, through just one entity, not only of the forces but embracing also the matter between which the forces act? I shall try to indicate how this might be achieved, but I apologize in advance because these ideas are difficult to express in qualitative terms.

To illustrate the point, consider again the Kaluza and Klein theory which unites gravity and electromagnetism, or gravitons (the quanta of the gravitational field) and photons, the mediators of these forces, as *one* entity representing curvature in a five-dimensional space (one of whose dimensions gets 'curled-up', or 'compactified' to use the technical expression). This is fine, but what is the electron, for instance, in this picture? Is it another entity, totally different from the entity representing the graviton and photon?

Now consider the actual situation at present. We have four forces: strong nuclear, weak nuclear, electromagnetic, and gravitational; and their mediators: coloured gluons, the $W^{\pm}$ and Z^{0}, the photon, and the graviton, all with a spin of 1 except the graviton, which is required to have spin 2 by Einstein's theory. Then we have matter, which is composed of quarks and leptons, or perhaps eight preons, all with spin 1/2. If *matter* as well as the *forces* are all to be represented by a single entity, then we must look for yet another over-all symmetry principle which can encompass these diverse spins of the *force mediators*, on the one hand, and leptons and quarks, or preons, the *constituents of matter*, on the other. In the world we know this symmetry must be well hidden, spontaneously broken, but in some very high-energy regime where the symmetry holds we conjecture that mediators of the basic forces (carrying integar values of spin) would appear as components of a fundamental entity, of which leptonic, quark, or possibly preonic matter (carrying half-integer spin) would form yet other components.

This brings us to some recent and exciting developments in the search for a unified theory encompassing the four forces, their mediators and matter. The first step was the discovery of a quite new type of symmetry, 'super-symmetry' which brought integer-spin and half-integer spin particles together within a single 'multiplet'. Hitherto, such states had been treated independently, now, within super-symmetry, they could be handled as equivalents. Amazingly, the next step, to introduce a gauge invariance, immediately leads to the appearance of the gravitational force as the gauge force corresponding to this symmetry. The new theory is called 'super-gravity'; it includes the spin 2 graviton and also, as gauge objects, particles of spin 3/2 called 'gravitinos'.

Now we need to bring in the electro-nuclear charges and combine spinning-matter not just with gravity but with all forces. An attempt has been made to extend super-gravity by combining super-symmetry with an internal symmetry, called S0(8), and requiring gauge invariance. The result, called 'N = 8 super-gravity', possesses a unique entity describing mediators of forces as well as matter in a single multiplet of possible physical states classified according to their spins as follows:

Spin	Number of states
2	1
3/2	8
1	28
1/2	56
0	70

This theory, if correct, would accomplish all unifications, of all forces and all matter. We have here the spin-2 graviton and 8 gravitinos of spin 3/2; 28 particles of spin 1; 56 particles of spin 1/2 and 70 spin-less states. But, the 28 spin-1 states unfortunately cannot include the W^+ and W^- of the electro-weak interaction and neither do the 56 spin-1/2 states include all the leptons and quarks. The spin-0 states might include the Higgs particle required for symmetry breaking and the generation of mass. But without the W^+, W^-, muon, and t-quark the scheme does not appear to match our needs.

However, perhaps we were wrong in attempting to identify this multiplet with known quarks, leptons, and force mediators. Perhaps we should go to the preon stage. Cremmer and Julia have shown that a formulation of super-gravity in 11 dimensions of space–time is equivalent, in 4 dimensions, to an extended super-gravity which has, as an unexpected bonus, an enlarged internal symmetry among the 8 electro-nuclear charges. The additional 7 dimensions of space-time are assumed to be 'curled-up', hidden from view in the first 10^{-43} seconds, like the fifth dimension of Kaluza and Klein. The enlarged symmetry can accommodate more states but the really important point is that the theory is unique: it cannot accommodate more than 8 charges plus gravity; the 8 spin-3/2 objects described by the theory could be the basic preons, each carrying one of the 8 charges. *If this theory is correct, we may be very near to a final, complete unification of all forces with spinning-matter, and with the fundamental charges being manifestations of hidden dimensions of space!*

But how can we test these ideas? Scherk has made a remarkable suggestion in this connection. The spin-1 states of extended super-gravity include 'gravi-photons' which raise the dramatic possibility of a repulsive, 'anti-gravity' component of force between matter which in some circumstances might cancel the normal, attractive gravitational force. This, in Scherk's view, might lead to small departures from the present laws of gravity which could have remained undetected for matter separated between 100 metres and 1000 metres; this sounds like science fiction! But is it so crazy? After all, the idea of creating, in the laboratory, the $W^\pm$ and Z^0 with masses nearly 100 GeV would have seemed crazy to us in 1958.

Summary

Let me now summarize our discussion of the stages of unification. The unified gauge theory of the weak nuclear force and electromagnetism led us to the prediction of a new force, a weak force mediated by a new spin-1 particle, the Z^0. This weak force has been observed, it behaves exactly as expected; now we await the discovery of the Z^0 and its charged partners, the W^+ and W^-. These are predicted to have masses of 90 and 80 GeV and above these energies the strengths of the two components of the electro-weak force become comparable.

Experiments at these energies and above, in accelerators such as LEP, are necessary not only to observe the Z^0 and $W^{\pm}$ but also for a thorough study of their behaviour to test the detailed predictions of the electro-weak gauge theory. Another essential component of the scheme is the Higgs particle, the spin-0 boson responsible for breaking the symmetry between the weak and electromagnetic interactions and introducing the mass differences between photons and the Z^0, $W^{\pm}$. At least one Higgs particle ought to exist, but the theory does not predict its mass.

Unification of the electro-weak and the strong nuclear force leads to the prediction of an unstable proton and so a Universe eventually depleted of protons and neutrons. The latter we shall not live to see, but the experimental task of detecting proton decay is now being enthusiastically tackled by experimental physicists. The present limit of proton life is 10^{30} years and it seems possible to conceive experiments which could detect a lifetime if it is less than 10^{32} years, as expected in present theory. There is a question of what the unification energy for the electro-weak and the strong nuclear force is likely to be. One school of theoreticians believes that the unification energy, where the strengths of the electro-weak and strong forces become comparable, is 10^{15} GeV, quite beyond the reach of any experiments. In this view we can never expect to observe directly the massive electro-nuclear bosons, the lepto-quarks, responsible for turning quarks into leptons; their existence could be inferred only indirectly through proton decay. There may indeed be, above energies of 100 GeV, a barren plateau in physics devoid of interest up to energies of 10^{15} GeV!

There are others who contest this point of view. If quarks and leptons are composite, a substructure may show up at much lower, though still high, energies. Pre-quarks or preons, perhaps carrying magnetic monopole charges, may emerge into view above 10^5 GeV (not an unreachable energy?). These may have a natural place in the ultimate unification, with gravity. This final step of unification may unite not only the four forces and their mediators, but these with spinning-matter as well. Hall-marks of this unification would be the existence of spin 3/2 gravitinos and anti-gravity.

This last unification of gravity with the electro-nuclear forces and with matter would take place around energies of the order of 10^{19} GeV, the so-called Planck energy. But if we have to test our theories at these energies of 10^{15} GeV and 10^{19} GeV, the only laboratory conceivable is the Universe itself, during the first instants of the Big Bang!

To conclude, I am always amazed by what we find at each successive level we explore. I should like to quote a prediction made by J. R. Oppenheimer twenty-five years ago which has been amply fulfilled in a way he could never have dreamed of in the unification ideas we have been talking about:

> Physics will change even more. . . . If it is radical and unfamiliar (now) . . . we think that the future will be only more radical and not less, only more strange and not more familiar, and that it will have its own new insights for the inquiring human spirit.

6

The Very Large and the Very Small

JOHN ELLIS

The physical world about us is governed by the set of four fundamental forces: gravitation, electromagnetism, the weak and the strong nuclear forces. A very deep understanding of these forces is not usually needed to describe everyday phenomena in the relatively peaceful and quiet part of the Universe close to us. Thus the motions of falling apples, stars, galaxies, and even the Universe in the large can be understood without knowing how to quantize gravity. Chemical and solid-state phenomena can be adequately described by quantum electrodynamics with values of the fine structure constant ($\alpha = 1/137$) and particle masses put in by hand; one does not need to know why they take the values they do. Similarly the gross features of the Sun's output of energy and its lifetime can be understood in terms of a few God-given parameters of the weak and strong nuclear interactions. We do not need to know the origin of these parameters, although stellar evolution would have been grotesquely different and human life totally impossible if some of them had taken even slightly different values.

In recent years great strides have been made in the understanding of the fundamental forces and in their unification into a simpler and more complete framework. Among the insights gained from such unified theories is increased understanding of the values of some of the previously arbitrary fundamental constants. For example the grand unification of strong, weak, and electromagnetic interactions entails that the fine structure constant be small. These insights establish an indirect connection between modern theories of matter and everyday phenomena in the Universe about us. If we are to look for more direct manifestations of our new unified theories we must look farther afield than our own uneventful neighbourhood in a nondescript galaxy. My purpose in this chapter is to describe how the behaviour of the Universe on a large scale in time and space is connected with deep aspects of the fundamental interactions that are not directly discernible in our everyday world. Indeed we now see ever more clearly that modern ideas in particle physics exist in intimate symbiosis with some fundamental aspects of cosmology and astrophysics, such as the very early history of the Universe and the origin of matter.

It is not difficult to understand why the Universe is an ideal arena for studying fundamental physics. One reason is quite simply its large size, which provides us with a long 'lever-arm' for performing delicate balance experiments or, if you prefer, a gargantuan telescope for magnifying the very small effects of individual particles. For example, the Universe, being large, contains many neutrinos, which it can 'weigh' much more precisely than laboratory experiments working with individual neutrinos. A second reason why the Universe is a good high-energy physics laboratory is that very violent and energetic events occur 'out there' in conditions very difficult or impossible to reproduce artificially on our own tranquil planet. For example, while the highest energies presently attainable with our particle accelerators are less than 10^3 GeV, it is believed that particle energies of 10^{19} GeV may be reached in black hole explosions, and were probably routine at very early stages in the Big Bang.

In order to show how we can utilize these potentialities of the Universe as a high-energy physics laboratory it will first of all be necessary to recapitulate some basic features of the 'standard' cosmological model, starting off from the Big Bang. This model is motivated by the observed over-all expansion of the Universe today, by the discovery of the relic 3 K microwave radiation, and by the qualitative success of calculations of the primordial abundance of helium. The success of these nucleosynthesis calculations and the continuing expansion of the Universe impose limits on the number and masses of neutrinos that are much more restrictive than those found in laboratory experiments. Additional information on the nature of the neutrinos may come from experiments searching for the flux of neutrinos emanating from nuclear reactions in the Sun. However, the most exciting interface between cosmology and high-energy physics may go back much earlier in time, to an epoch when the Universe was very compressed and typical particle energies were of order 10^{15} GeV. It now seems likely that an explanation of the origin of all the matter in the Universe may be found in the grand unification of strong, weak, and electromagnetic particle interactions at these exceedingly high energies. These grand unified theories also provide a melancholic vision of the ultimate future death of matter in the Universe.

The standard cosmological model

Observational cosmologist are agreed that the Universe is expanding with the galaxies flying apart from one another. While they may disagree on the rate of expansion, they believe that it is the relic of a primordial explosion, the Big Bang, which took place about one or two times 10^{10} years ago. The gross features of the evolutionary history of the Universe in this standard cosmological model are summarized in Table 6.1. Shown there are the approximate times at which various important evolutionary milestones are believed to have been passed, the approximate temperature of the Universe at each of these epochs, and the corresponding average energies of its constituent particles.

The best understood epoch in the Table is obviously the present one when the Universe is about 10^{10} years old. Most of the visible matter is now collected in stars which shine by 'burning' lighter elements in order to make heavier ones using reactions like:

$$p + p \rightarrow D + e^+ + \nu_e, \qquad p + e^- + p \rightarrow D + \nu_e,$$
$$p + D \rightarrow {}^3He + \gamma, \qquad {}^3He + {}^3He \rightarrow {}^4He + 2p + \gamma,$$
$${}^7Be + e^- \rightarrow {}^7Li + \nu_e, \text{ etc.}$$

which contribute to the production of energy in our Sun. These reactions emit many electron-neutrinos, ν_e, and have generally been considered to be well understood since the work of Bethe, Hoyle, and others. Most of the visible stars are collected into galaxies which are in turn gathered into clusters. It is believed that our local cluster of galaxies contains no substantial concentration of antimatter, but only conventional matter. If our cluster contained antimatter as well, then one would expect there to be regions where the matter and antimatter would meet and annihilate, creating high-energy photons (γ-rays and X-rays) for which no evidence has ever been seen. Some antiprotons have recently been seen in high-energy cosmic rays, but their flux is low enough for them to be explained as the secondary

Table 6.1

Time	*Event*	*Temperature*	*Typical energy*
10^{-45} seconds	?—Quantum gravity effects are strong	10^{32} K	10^{19} GeV
10^{-35} seconds	Quarkosynthesis—the predominance of matter established?	10^{27} K	10^{14} GeV
100 seconds	Nucleosynthesis—helium and deuterium created	10^{9} K	10^{-4} GeV = 1/10 MeV
10^{6} years	Photon decoupling—the background radiation originates	10^{3} K	10^{-10} GeV = 1/10 eV
10^{10} years (Now)	Galaxies, stars, and we exist	3 K	10^{-3} to 10^{-4} eV
		for background radiation, but no uniform temperature throughout the Universe	
$10^{31\pm2}$ years	All matter decays?		

products of collisions involving primary nucleons, not antinucleons. It seems likely that galaxies and galactic clusters may contain invisible, or 'dark', matter in addition to that evident in stars. The existence of this matter is inferred from its indirect gravitational effects, and it could take many forms, such as mini black holes, Jupiter-sized planets, dust clouds, or even neutrinos.

The different galactic clusters are receding from one another at a rate of about 20 kilometres per second for each million light-years by which they are presently separated. A natural question is whether there is sufficient matter in the Universe for its gravitational attraction to cause the Universe to collapse back on itself eventually. This would require the present matter density to exceed about 2×10^{-29} gm per cubic centimetre, and current belief is that the matter density is somewhat less than this critical density; in which case we expect the Universe to continue expanding indefinitely far into the future. Near the end of this chapter we shall return to this distant future; for the moment let us restrict ourselves to the extrapolation of the present expansion back into the past. This extrapolation is relatively simple and reliable because the Universe appears to be rather homogeneous and isotropic on a large scale. Of course, local inhomogeneities exist, we are part of the evidence, but on a larger scale the distribution of galactic clusters looks similar in all directions from whatever point they are viewed.

A naive extrapolation backwards in time of this present symmetric expansion suggests that there is a singularity in our past. Matter would have been very compressed close to this putative singularity, and we are familiar with the idea that matter heats up when it is compressed. Thus, as we go further and further backwards in time we expect matter to have been at higher and higher temperatures. At an early stage in this Big Bang matter would have been in thermal equilibrium with a 'bath' of hot radiation, photons.

Now, starting from such a situation, let us consider what would happen as the Universe evolved. As the Universe expanded and cooled radiation and matter would eventually have *decoupled*, thereafter cooling down separately. A few words of explanation are due here. In a system in thermal equilibrium the proportions of total energy carried by the different components, matter and radiation in this case, remain constant. This requires continuing interactions between the components in order that there can be a balance between the rates of absorption and of emission of radiation by the matter present. Expansion and cooling can affect this balance by causing the collisions to become less frequent, and at low temperatures the typical collision energies can fall below the threshold energies necessary for some interactions to proceed. At the time of decoupling, the reactions between radiation and matter are no longer able to maintain an equilibrium sharing of energy and so the two components no longer share a common temperature.

Matter and radiation would have decoupled at a temperature a little above 1000 K, when the Universe was almost a million years old. The photons were no longer energetic enough to disrupt atoms and there were no more free electrons to collide with. After this, the radiation would have continued to cool according to the simple formula:

$$\frac{T_{\text{now}}}{T_{\text{decoupling}}} = \frac{R_{\text{decoupling}}}{R_{\text{now}}},$$

where T is the radiation temperature and R the characteristic size of the Universe at any given time. Calculations indicate that $(R_{\text{decoupling}}/R_{\text{now}})$ should be about 10^{-3}, and so any radiation that is a relic of the Big Bang should now have a temperature of a few kelvins. Lo and behold, just such a background radiation was found in 1964 to 1965 by Penzias and Wilson (Fig. 6.1) with an apparent temperature of about 3 K, just as expected in the Big Bang model. Subsequent observations have confirmed the expected 'black-body spectrum' for the distribution in temperature of this radiation, and found that it is isotropic to better than one part in 10^3. Its existence is considered one of the two greatest successes of the Big Bang model.

Because of this black-body radiation, the Universe now contains very many photons, about 300 per cubic centimetre. However, there seem to be relatively few baryons, that is neutrons and protons, in the Universe as a whole, their density being of order 10^{-7} per cubic centimetre. This means that the ratio of the numbers of baryons to photons is very small: N_B/N_γ is between 10^{-10} and 10^{-9}. Before the advent of grand unified theories of particle interactions this number was either puzzled over or ignored. If there is no antimatter in the Universe and never was, why is N_B/N_γ so small, and not of order unity? On the other hand, if the primordial 'soup' had started off symmetric, with equal amounts of matter and antimatter, surely they would *all* have annihilated to form many more photons than the 'small excess' of 10^9 or so indicated above? We shall return later to the possible resolution of this puzzle; for the meantime let us resume our rearward odyssey into the Charybdis of the Big Bang and encounter its second great success.

The previous milestone in the evolution of the Universe shown in Table 6.1 was when the temperature was in the neighbourhood of 10^{10} K, and hence particle energies were about the 1 MeV characteristic of both the neutron–proton mass difference and the mass of the electron. When the temperature was somewhat higher than this there would have been essentially equal numbers of neutrons and protons; and the light particles, electrons, positrons, neutrinos, and photons, would all have been in thermal equilibrium.

As the temperature dropped below 10^{10} K the ratio of neutrons to protons plus neutrons (the mass fraction of neutrons) would have fallen below 0.5; this is because the relative numbers of neutrons and protons in a state of thermal equilibrium would be determined by a Boltzmann factor: exp $(-\Delta M/T)$, where ΔM is the neutron–proton mass difference, 1.3 MeV, and T is the temperature (expressed in equivalent energy units). This is the

FIG. 6.1. Wilson and Penzias in front of the 20-foot horn antenna which detected the 3 K cosmic photons, relics of the Big Bang. (Bell Telephone Laboratories.)

condition illustrated at time 10 seconds, T about 3.10^9 K, in Fig. 6.2. The decline in number of neutrons, at this stage, is mainly due to the process:

$$\nu_e + n \rightarrow p + e^-$$

and, somewhat later, neutron decay:

$$n \rightarrow p + e^- + \bar{\nu}_e,$$

now unbalanced by the reverse reactions like:

$$p + e^- \rightarrow n + \nu_e,$$

which requires energies greater than 1 MeV.

More than 100 seconds after the Big Bang, when the temperature had fallen below 10^9 K, the thermal energies would have been sufficiently low for any nuclei formed to avoid destruction. First deuterium and from that copious amounts of helium would have been produced. Helium is more tightly bound than the other light nuclei and for this reason it is by far the most abundant of the nuclei made at this stage.

In this process essentially all the remaining neutrons would have been used up and since helium contains two neutrons and two protons the helium mass faction, Y_{He}, would have reached a value close to twice the neutron fraction at the start of nucleosynthesis (see Fig. 6.2). The neutron fraction at this time is in fact rather sensitive to the precise rate of expansion of the Universe, which is in turn controlled by the density of matter and the number of light particles with masses less than 1 MeV, such as neutrinos. We shall return later to this sensitivity to the number of neutrinos. For the

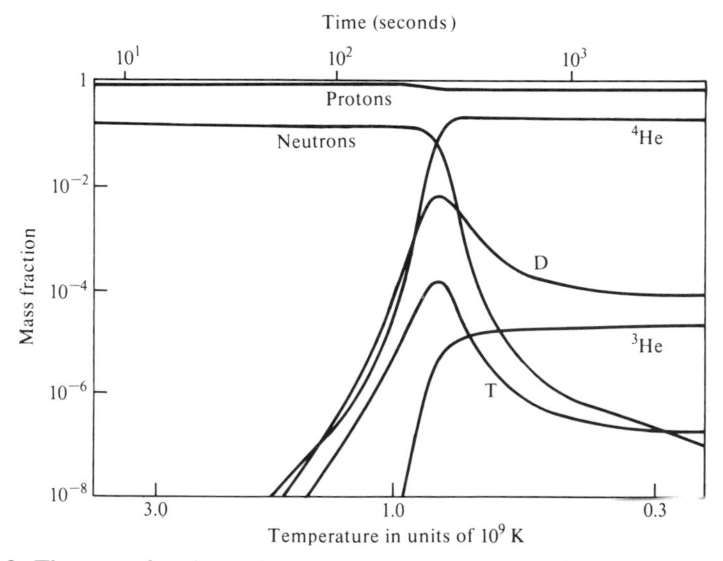

FIG. 6.2. The mass fractions of neutrons, protons, and the light nuclei helium-4, helium-3, tritium, and deuterium from a calculation of primordial nucleosynthesis.

moment we content ourselves by noting that these calculations of primordial nucleosynthesis have yielded a helium mass faction Y_{He} of order 25 per cent, and that this is qualitatively consistent with most present astrophysical observations, once the subsequent manufacture of a small amount of helium by first generation stars is subtracted out.

This primordial nucleosynthesis would have been very inefficient for making nuclei heavier than helium, because of the famous 'gap' due to the absence of stable nuclei containing a total of five neutrons and protons. As there is ample evidence of the existence of heavier nuclei it is necessary to add that quite different conditions apply in the centres of stars, so that in time the gap is bridged, leading to the creation of heavier nuclei. These can then be distributed by stellar explosion to build the familiar world about us.

The prediction of the helium fraction is the second great success of the Big Bang model and suggests that it is legitimate to extrapolate our present expanding and cooling, homogeneous and isotropic Universe back to an epoch when it was 10^{15} times younger and 10^{11} times hotter than it is today. What happened still earlier? We know of no reason why a naive extrapolation back to yet earlier times should not be valid. On the other hand we have no evidence that such confidence is in fact justified. Nevertheless, later in this chapter we shall extrapolate back to an 'age' 10^{38} times earlier still, when the temperature was 10^{18} times higher even than at nucleosynthesis. But before indulging in this hubris we shall first see what the relatively well-understood epochs of the Big Bang can tell us about particle physics, and about neutrinos in particular.

Neutrinos in particle physics

Neutrinos are very light, electrically neutral particles, which appear to have only weak interactions. They seem to come in at least three varieties, each one associated with a different charged lepton as indicated:

$$\begin{pmatrix} \nu_e \\ e^- \end{pmatrix}, \quad \begin{pmatrix} \nu_\mu \\ \mu^- \end{pmatrix}, \quad \begin{pmatrix} \nu_\tau \\ \tau^- \end{pmatrix}, \quad \ldots ?$$

Laboratory experiments have established upper limits on their masses that are of varying stringency. From the decay of tritium,

$$^3\text{H} \rightarrow {}^3\text{He} + e^- + \bar{\nu}_e,$$

we know that the mass of the electron-neutrino, $m(\nu_e)$, is less than about 10^{-4} of the electron mass. Similarly from muon decay,

$$\mu^- \rightarrow e^- + \bar{\nu}_e + \nu_\mu,$$

we know that $m(\nu_\mu)$ is less than 10^{-2} of the muon mass and from

$$\tau^- \rightarrow (3\pi)^- + \nu_\tau$$

we know that $m(\nu_\tau)$ is less than one tenth the τ-lepton mass. There is no fundamental reason known why the neutrino masses should be strictly zero, and our present prejudices indeed suggest that they should not be. In other instances Nature associated the masslessness of a particle with an exact local conservation law or symmetry. For example, the masslessness of the photon is believed to be associated with the local conservation of electric charge, which is related to a gauge symmetry of the electromagnetic force. There is no known local charge which could be similarly associated with the masslessness of the neutrinos. Modern grand unified theories of the fundamental interactions go further and suggest that the neutrino masses may in fact be of order 1 eV, give or take a few orders of magnitude!

Another uncertainty about neutrinos concerns the total number of different species. We have direct evidence for two, the ν_e and ν_μ, and indirect evidence for the third, the ν_τ. Any other would presumably be associated with leptons heavier than the τ, which we have not yet been able to detect in our laboratories. There are very indirect indications from the strength of neutral weak interactions that there cannot be more than about a thousand such very heavy leptons. A more direct limit on the total number of neutrinos comes from the fact that nobody has ever seen a decay of the type $K^\pm \rightarrow \pi^\pm$ + nothing. If we interpret the 'nothing' as a sum of all possible types of invisible neutrinos with masses less than $(m_K - m_\pi)/2$, then we get an upper limit of about 6000 on the total number of neutrino types! We know of no better limit on the number of neutrinos coming from present-day laboratory experiments on particle physics, though big improvements are possible with new particle accelerators now being planned (e.g. LEP, see chapter 7). However, we shall see in a moment that cosmology can already give us much more stringent limits on the number of varieties of neutrinos.

Astrophysical and cosmological restrictions on neutrinos

The first connection between the very large and the very small comes from an aspect of astrophysics that we think we understand very well, namely the Sun. As mentioned earlier, the Sun is believed to shine because of nuclear interactions in its interior, and these processes create many electron-neutrinos. An experiment has been set up to look for these neutrinos via interactions of the form $\nu_e + {}^{37}\text{Cl} \rightarrow e^- + {}^{36}\text{Ar}$. After some initial doubts, it now seems clear that solar neutrinos are indeed being seen at a rate of (2.2 ± 0.3) Solar Neutrino Units (SNU); where one SNU corresponds to 10^{-36} argon atoms produced per second per chlorine atom. The number observed is rather lower than the number expected, which is about 5 to 9 SNU, although the calculations are very difficult to make precise. Various mechanisms have been proposed to explain away the possible discrepancy. One possibility is that some major or minor detail of the conventional nuclear physics calculations of the Sun's interior has gone awry. It should be pointed out that the solar neutrino experiment has only been sensitive to a small component high in the distribution of neutrino energies expected to come from the Sun. The neutrinos detected arise from the $^7Be \rightarrow {}^7Li$ reaction mentioned above; this is rather peripheral to the main energy-generating processes and may well have been miscalculated without our ideas about the central energy-producing reactions in the Sun being wrong. But suppose the observations and the stellar models are both correct, what could then be the explanation of the solar neutrino puzzle?

One possibility is that, after all, the neutrinos do have masses. Then a consequence of quantum mechanics can lead to rather bizarre effects. The neutrino states with definite mass need not be the particular forms of neutrino (ν_e, ν_μ, or ν_τ) observed in laboratory experiments to be produced in association with the charged leptons (e, μ, or τ, respectively). In this case the neutrinos, ν_e, produced in the Sun are to be described as a linear combination

$$\nu_e \equiv \alpha_1\nu_1 + \alpha_2\nu_2 + \alpha_3\nu_e + \ldots$$

of the neutrino states ν_1, ν_2, ν_3 . . . which do have definite masses. The quantities α_1, α_2, α_3 . . . are the amplitudes of each type composing the wave which describes the propagation of the neutrinos. As the neutrinos from the Sun journey through space, components produced with the same energy will propagate with different velocities because of their different masses. This means that our initial neutrino wave

$$\alpha_1\nu_1 + \alpha_2\nu_2 + \ldots$$

will arrive at the Earth as a different linear combination

$$\beta_1\nu_1 + \beta_2\nu_2 + \ldots.$$

This will in general correspond to a combination

$$a_e\nu_e + a_\mu\nu_\mu + a_\tau\nu_\tau + \ldots$$

of neutrinos which is not a pure ν_e state. In this case the number of ν_e arriving at the Earth will be reduced by a factor of $|a_e|^2$ compared with that expected, and there will be the same reduction in the number of SNU observed:

$$\frac{\text{SNU seen}}{\text{SNU expected}} = |a_e|^2 < 1.$$

For this explanation of the reduced number of SNU observed to be valid, it is necessary that the mass differences between the different states ν_i not be too small, in fact greater than about 10^{-5} eV. This is consistent with the limits coming from laboratory experiments as well as the more stringent cosmological constraints on neutrino masses to be discussed below. However, it is still not clear that the discrepancy between the observed and expected number of SNU is really serious, and many other explanations are still possible besides that of massive neutrinos.

If neutrinos really do have masses, there is another interesting astrophysical method to 'weigh' them up using the large scales afforded by the Universe. The point is that when a supernova collapses and dies, many neutrinos are emitted. Since the different mass states would propagate towards us at different velocities, it is possible that if such a collapse occurred many light-years away, there would be measurable time delays between the arrival times of the pulses of different types of neutrinos. Some of the large underground experiments now being built to look for the decays of protons, expected in some grand unified theories of particle interactions, may also be able to detect these staggered death-throes of distant supernovae.

In the meantime, some cosmological considerations already give us strong restrictions on the masses and numbers of neutrinos. These have their origins in the discussion of nucleosynthesis in the standard Big Bang model. Another development as the temperature dropped below 10^{10} K, where particle energies were typically about 1 MeV, would have been the decoupling of neutrinos from electrons and positrons as the thermal energies would have become insufficient for the reaction $\nu + \bar{\nu} \rightarrow e^+ + e^-$ to proceed; then the inverse reaction, $e^+ + e^- \rightarrow \nu + \bar{\nu}$, would have allowed all species of neutrinos with masses much less than 1 MeV to be produced in large and essentially equal numbers. Analogously to the photons, which decoupled later and have been observed by Penzias and Wilson, we must now be bathed in an invisible 'sea' of about 100 neutrinos of each type per cubic centimetre, with a temperature of about 2 K and corresponding thermal energies of order 10^{-4} eV. All except the very lightest (mass $<10^{-4}$ eV) of such neutrinos would now have velocities much less than the velocity of light and so behave according to ordinary Newtonian mechanics. They would therefore contribute to the present matter density of the Universe simply in proportion to the sum of their masses. At the moment the Universe does not seem to be on the verge of collapsing under its own gravitational attraction, which gives us an upper limit on its mass density. This upper limit on the present density of the Universe in turn implies an upper limit of about 100 eV on the sum of the masses of all types of neutrino much lighter than 1 MeV. Since we already know from laboratory experiments that the electron and muon neutrinos are lighter than 1 MeV, this cosmological limit applies directly to them. If, for example, there are no more than three types of neutrino with mass less than 1 MeV and if all have the same mass (there is no reason why they should) then this result places an upper limit of about 30 eV on this mass. Even neutrinos, or other heavy neutral leptons, with higher masses are restricted by cosmology, as long as they are stable. We see from Fig. 6.3 that the mass density of such heavy leptons would cause the Universe to collapse gravitationally if the masses were between about 100 eV and a few GeV. To establish these limits one must believe in the standard Big Bang cosmological model back somewhat before the epoch of nucleosynthesis, but this extrapolation is relatively conservative compared with those we will be making in the next section.

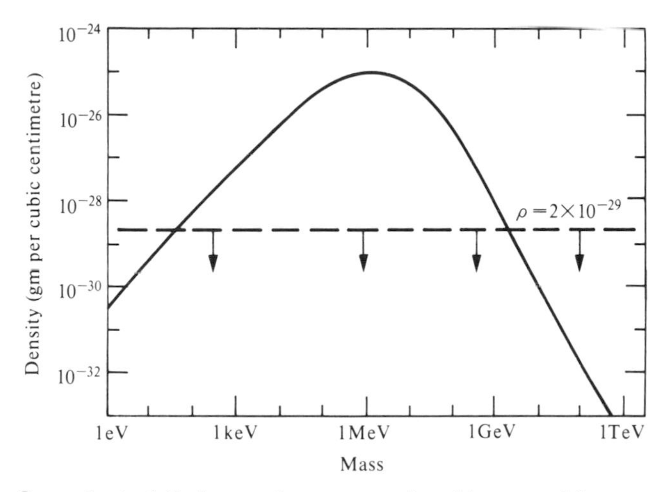

FIG. 6.3. Cosmological limits on the masses of stable neutral leptons or 'heavy' neutrinos. If there were any such objects with masses between about 100 eV and a few GeV their mass density would be higher than the critical density (about 2×10^{-29} gm/cm^2) for continuing expansion, and the Universe would eventually collapse under its own gravitational attraction.

The Big Bang model can also be used to make arguments against heavy stable neutral hadrons as well as neutral leptons. In this case, however, we do not use the relatively weak constraint imposed by the present matter density of the Universe. Instead we use the fact that such heavy stable hadrons would form unusual heavy isotopes of common elements such as oxygen. Stringent upper limits have been placed on the abundances of such heavy isotopes which argue against stable neutral hadrons with masses up to about 40 GeV. For comparison, the best limit to date on stable hadrons produced by high-energy particle accelerators only restricts them to masses above about 15 GeV.

We have seen that cosmology establishes limits on the possible masses of elementary particles which are much more restrictive than those established with particle accelerators. Another very dramatic limit from cosmology is that placed on the total number of different neutrino types. Remember that neutrinos with masses much less than 1 MeV are very copious around the time of nucleosynthesis. They make a substantial contribution to the total energy density and pressure due to elementary particles in the early Universe, and thereby affect its expansion rate. The faster the expansion, the faster the cooling and the sooner conditions favourable to building up helium are reached. Also there will have been less time for neutrons to decay away and so more helium can be made. Numerically, the effect of adding one neutrino type, with its associated antineutrino, to the primordial 'soup' is to increase the helium mass fraction by about 1 per cent. As we have seen, the best value for the primordial helium fraction, after allowing for subsequent stellar manufacture, is close to 25 per cent. This restricts the number of different neutrino types to about 3 or 4, as seen in Fig. 6.4. Compared with the laboratory limit of several thousand types this is a considerable improvement, though it should be remembered that this limit only applies to neutrinos weighing considerably less than 1 MeV.

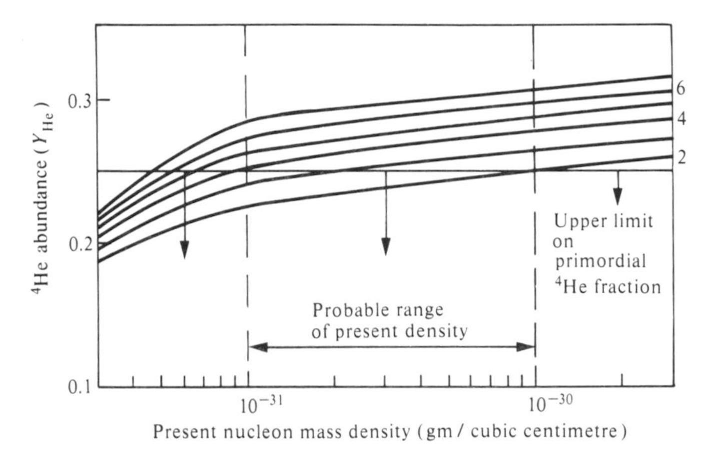

FIG. 6.4. The sensitivity of the present helium abundance, Y_{He}, to the number of 'light' neutrinos with masses much less than about 1 MeV. The upper limit of 25 per cent on the primordial value of Y_{He} suggests that there are at most three or four pairs of 'light' neutrinos and their antineutrinos.

The origin of matter

We saw in the previous section that astrophysics and cosmology can be of great use to particle physics in restricting the possible numbers and masses of elementary particles. In this section we shall see how grand unified theories of particle interactions may return the compliment by explaining a long-standing cosmological puzzle, the origin of matter. Previous applications of nuclear physics have shown how the heavy elements observed in Nature can have been created out of the lighter elements in stars. Going further back in the history of the Universe, the creation of the lighter elements, helium and deuterium, has been explained by primordial nucleosynthesis in the Big Bang. The remaining problem is the origin of the neutrons and protons used in this synthesis, with the challenge of understanding why we only see matter and no antimatter in our local cluster of galaxies, and by extrapolation in all the visible Universe.

No-one was ever able to give a satisfactory explanation of these mysteries in terms of the conventional electromagnetic, weak and strong nuclear forces seen and studied at low energies. Thus to seek an explanation we must go on to higher energies corresponding to higher temperatures and hence earlier eras in the expansion of the Universe.

The experiments on lepton scattering on protons and neutrons described by Donald Perkins in chapter 4 suggest that at very high energies hadronic particles act as if they are made of almost free constituents called quarks. We therefore analyse the hot 'soup' that was the very early Universe in terms of quarks and leptons and the relatively feeble interactions between them. The problem of the present dominance of matter over antimatter then becomes the problem of understanding why the density of quarks in the primordial 'soup' should have been greater than the density of antiquarks. Calculations indicate that if this trick could be realized in some way, then essentially all the antiquarks would have combined with quarks when the temperature fell to about 10^{13} K, corresponding to an energy of 1 GeV, leading indirectly, through the formation of mesons, to the creation of photons comparable in number to the present photon number. Thus, if the supposed primordial excess of quarks had been about 10^{-9} to 10^{-10} of the total number of quarks and antiquarks, then a similar fraction of quarks would have been left over after all the antiquarks paired off with quarks, rather like the wallflowers at a dance. These wallflower quarks would have grouped together into protons and neutrons and would survive today, giving the observed ratio of matter density to photon density:

$$N_B/N_\gamma \approx 10^{-9} \text{ to } 10^{-10}.$$

So how do we establish a net quark asymmetry of about 10^{-9}? The explanations proposed in the 'bad old days' included starting the Universe off with only quarks and generating 10^9 times more quark–antiquark pairs by mysterious dissipative phenomena. Another response was simply to abandon the problem by postulating a small primeval quark asymmetry when the Big Bang started. A much more aesthetic initial condition would have been to start with equal numbers of quarks and antiquarks, and hence zero net baryon number. In order to generate the observed non-zero net quark or baryon density, it is therefore necessary to postulate interactions that change baryon number. Although no such interactions have ever been seen there is no sacred law preventing their existence, because there exists no massless particle which could be associated with a law of baryon conservation. Indeed, it is commonly believed that baryon number is not conserved in the strong gravitational fields associated with black holes.

As discussed by Abdus Salam in chapter 5, interactions changing baryon number are the rule rather than the exception in grand unified theories of particle interactions. The simplest example of such a transition is the annihilation of two quarks to make an antiquark and an antilepton:

$$\mathrm{q} + \mathrm{q} \rightarrow \bar{\mathrm{l}} + \bar{\mathrm{q}},$$

which changes the net quark number by -3 and hence reduces baryon number by 1 (since baryon number is 1/3 for quarks and $-1/3$ for antiquarks). Such interactions should cause protons and bound neutrons to decay via modes like

$$\mathrm{p} \rightarrow \mathrm{e}^+ + \pi^0 \text{ or } \mathrm{n} \rightarrow \mathrm{e}^+ + \pi^-.$$

So far laboratory experiments have not confirmed that such decays occur, and have instead established a lower bound on the nucleon lifetime of about 10^{30} years. The lifetime expected for these decays is very sensitive to the mass of the heavy, spin 1 particle X, the lepto-quark, which is expected to be the carrier ($q + q \rightarrow X \rightarrow \bar{l} + \bar{q}$) of the 'hyperweak' baryon-number-changing interaction. The lower limit on the nucleon lifetime gives a lower bound on the mass of the X particle of order 10^{14} to 10^{15} GeV. This may seem unbelievably high when compared with the masses of other elementary particles, leptons, quarks, etc., but it is still much lower than the so-called Planck mass of order 10^{19} GeV at which the quantum effects associated with gravity probably become strong and unmanageable. In fact, when one tries to put the known fundamental interactions into a grand unified theory, one finds very naturally that the unification should occur at or slightly below an energy scale of 10^{15} GeV. The exchanges of particles with masses of this order should cause the proton to decay in 10^{33} years or less, and a number of experiments to look for protons decaying within this lifetime are now being prepared.

In order to exploit these possible baryon-number-changing interactions in the early Universe, we have to be foolhardy enough to extrapolate the standard cosmological model back to temperatures above 10^{27} K, corresponding to particle energies above 10^{14} GeV. It is only in this very early era, when the Universe was a mere 10^{-35} seconds from its birth, that one can expect a significant change in the net baryon density of the Universe to have taken place. However, in order for a net baryon asymmetry to have been generated, various properties must be satisfied by the baryon-number-changing interactions and the conditions in which they operate. Many of these requirements were enumerated by A. D. Sakharov in 1967, although his work was long before the presently fashionable grand unified theories came into existence.

The first requirement is that the baryon-number-changing interactions should violate the particle–antiparticle symmetry known as charge conjugation or C. Otherwise, starting from an initial state with equal numbers of quarks and antiquarks they will always generate equal numbers of quarks and antiquarks. Furthermore, the baryon-number-changing forces should not be left unchanged by the combined operation of changing particles into antiparticles and mirror reflection, known as CP. This transformation would change a particle moving in one direction into an antiparticle moving into the opposite direction. Hence, if it were an exact symmetry, the Universe would always contain equal numbers of quarks and of antiquarks if it had started out that way. The third requirement is that the evolution of the early Universe must possess an 'arrow-of-time'. If it did not, a perfectly general symmetry which is believed never to be violated, namely invariance under the combined operations of particle–antiparticle interchange, mirror reflection, and reversal in time (CPT), would leave the state of the Universe unchanged and guarantee equal numbers of quarks and antiquarks. The three properties of C (particle–antiparticle) asymmetry, CP (particle–antiparticle and mirror) asymmetry, and the existence of an 'arrow-of-time' are all believed to be present for grand unified particle interactions in the very early Universe.

We know that the familiar weak forces are not symmetric under either C or CP. In chapter 3 Chris Llewellyn Smith describes an experiment on the β-decay of cobalt-60 which demonstrates violation of mirror reflection symmetry by the weak force. Particle–antiparticle symmetry is also violated by the weak force, but symmetry under the combined operation, CP, was believed to be respected until in 1964 Cronin and Fitch and their colleagues discovered a very rare decay mode of the neutral K-meson which implied a subtle violation of this symmetry. When considering the decay of the neutral K-mesons one must take a linear combination of the K^0 and its antiparticle $\bar{K}^0$ (rather like the combinations of different neutrino types we met earlier). There are two such 'mixtures': one, $K^0{}_S$, has a short lifetime (about 10^{-10} s) and the other, $K^0{}_L$, has a lifetime 500 times longer (5×10^{-8} s). So far it is only in the decays of $K^0{}_L$ that the CP violating effects have shown up. One piece of evidence concerns the decay of $K^0{}_L$ to electrons; there are two possible modes:

$$K^0{}_L \rightarrow e^+ + \pi^- + \nu_e \text{ and } K^0{}_L \rightarrow e^- + \pi^+ + \bar{\nu}_e.$$

One of these two processes can be transformed into the other by the combined CP operation; if CP symmetry holds good, the rates for these two processes must be exactly equal. But they are not, the measured ratio is:

$$\frac{K^0{}_L \rightarrow e^+ + \pi^- + \nu_e}{K^0{}_L \rightarrow e^- + \pi^+ + \bar{\nu}_e} = 1.0067 \pm 0.0003.$$

A very small violation of CP symmetry is revealed by this small asymmetry in the numbers of electrons and their antiparticles emerging from $K^0{}_L$ decay. We have every theoretical reason to expect the same properties, of C and CP violation, to be true for the forces changing quark or baryon number.

The required 'arrow-of-time' is provided by the expansion of the Universe. As it expands it cools, and the interactions changing the net quark number become too slow to keep up with the expansion rate, so that they are no longer in thermal equilibrium. Then any quark–antiquark asymmetry that is generated by C- and CP-asymmetric forces cannot be 'boiled away' by other baryon-number-changing interactions.

Using these basic physical principles, a number of different models for generating a net baryon asymmetry have been proposed. In the simplest one to visualize, very early in the Universe when the temperature was very great there would have been many superheavy particles X and their antiparticles $\bar{X}$. The X particles could have decayed either into a pair of quarks or into an antiquark and an antilepton: $X \rightarrow q + q$ or $\bar{q} + \bar{l}$. Conversely, their antiparticles $\bar{X}$ could have decayed into pairs of antiquarks or into a quark and a lepton: $\bar{X} \to \bar{q} + \bar{q}$ or $q + l$. In the absence of particle–antiparticle symmetry and its combination with mirror symmetry, there is no reason why the fraction of X particles which decayed into pairs of quarks should be the same as the fraction of their antiparticles decaying into pairs of antiquarks. Under these circumstances, starting with equal numbers of particles and antiparticles, X and $\bar{X}$, we can finish up with unequal densities of quarks and antiquarks, as illustrated in Fig. 6.5.

Such a mechanism for generating a difference between the densities of

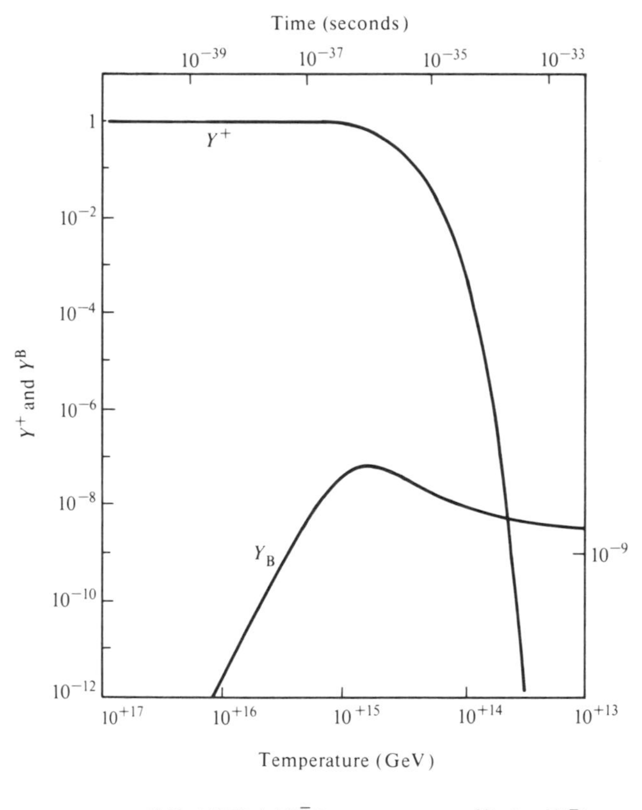

$$Y^+ = \frac{(1/2)(N(X)+N(\bar{X}))}{N(\gamma)} \quad ; \quad Y_B = \frac{N(q)-N(\bar{q})}{N(\gamma)}$$

FIG. 6.5. A calculation showing how by starting at high temperatures with equal densities of X particles and X antiparticles, their decays may ultimately generate an asymmetry between the densities of quarks and antiquarks. Y_+ is the sum of the densities of X and $\bar{X}$ and Y_B is the net quark–antiquark asymmetry measured relative to the number of photons, N_γ. At very early times, when particle energies were typically much greater than the X, $\bar{X}$ masses ($\sim 10^{15}$ GeV), the X and $\bar{X}$ densities were the same as that of the photons. Decays of X and $\bar{X}$ which violate quark conservation and the symmetries C and CP could lead to a quark excess when the X and $\bar{X}$ drop out of equilibrium, and eventually a baryon to photon ratio of about 10^{-9}.

quarks and antiquarks, and hence an eventual relative factor of 10^{-9} or less between the densities of baryons and photons today, would be a tremendous advance in our understanding of the origin of the matter of the Universe if its validity could be proved. Unfortunately, believing it requires an enormous extrapolation back to the remote origins of the Universe, and our evidence for its basis in grand unified theories, still just a theoretical speculation, can only be very indirect for the foreseeable future. If forces which change quarks into antiquarks and hence change the net number of baryons do exist, then we arrive at a rather melancholic vision of the future of our Universe. If, as seems likely, it does not contain a sufficient density of matter to cause it to collapse back on itself into a Big Crunch then the Universe will continue to expand forever. As it continues to expand, all the matter that it contains will gradually decay, perhaps within a time-scale of 10^{33} years. The only particles left will be the completely stable electrons, positrons, neutrinos, and photons. Competing with this disintegration will be a tendency for matter to gravitate together into black holes. However, even these black holes will eventually explode in bursts of radiation and particles. So the distant future of our Universe may be a diminishing miasma of disintegrating matter, punctuated by the occasional pyrotechnic black hole explosion.

Symbiosis between cosmology and particle physics

We have seen that there are many deep connections between modern cosmology and astrophysics on the one hand, and current developments in particle physics on the other, the very large and the very small. It is indeed a symbiotic relationship with each physical discipline aiding the other in the pursuit of its goals. Relatively well-understood aspects of cosmology and astrophysics provide the particle physicists with strong restrictions on the numbers and properties of the particles, especially neutrinos, that they are allowed to put into their theories. Conversely, some tried and tested aspects of particle physics, and some much more speculative ones, give us new ways of pondering some of the most profound cosmological problems that confront us, such as the origin of matter itself. But the daring of our speculations should not blind us to the incompleteness of our understanding and the modesty of our achievements.

7

The Tools of Particle Physics

J. B. ADAMS

It may seem odd that in a book dealing with a very fundamental branch of physics one chapter is devoted entirely to the tools that are used by the experimenters. In most scientific reviews such apparatus, if it is mentioned at all, is described briefly during the discussion of experimental results. In the present case the tools of research are given such prominence because they are very large, very expensive, and very few, and these characteristics have had a profound effect on the way this research has evolved over the past twenty years.

The basic requirements for experiments in high-energy particle physics are a particle accelerator, particle detectors, and data analysis systems. The purpose of the accelerator is to enable physicists to investigate processes taking place at the highest possible energies in order to reveal the small-scale structure of matter. The basic laws of quantum mechanics leave no option for an easier approach; an accelerator is, in a real sense, a microscope for sub-nuclear phenomena. The detectors, of various types, measure the properties and record the behaviour of the particles created at these high energies, most of which are ephemeral and do not exist naturally on Earth. The data analysis system, usually a chain of computers ranging from small, dedicated processors to very powerful central facilities, enables the physicist to reconstruct, from the raw data, a complete physical picture of the life and ultimate decay of all the particles taking part in the process under study.

Except for the computers, these tools cannot be bought from industry. They are designed by specialists employed by the big laboratories and by experimenters in universities, their components are manufactured by industry and they are assembled and commissioned by their designers who subsequently operate and develop them over many years. As their size and cost increased the particle accelerators became concentrated first in National Laboratories and then in International Laboratories, and the organization needed to carry out experimental research on the scale now typical is one of its most notable characteristics. Despite all this, however, the research itself remains a fundamental enquiry into the basic constituents of matter and the laws governing their interactions.

I shall describe in very broad outline the tools in use today in high-energy particle physics research, briefly discuss the problems which the size and cost of these tools have brought to the subject and end with some remarks on future possibilities.

Particle accelerators

Particle accelerators are by far the largest of the tools of the experimental physicist and the most costly. The biggest, of which there are only two in the World, measure about 2 km in diameter. Perhaps one of the most remarkable things about particle accelerators is the rapid rate at which they have grown in size over the last 50 years. The first were constructed in the early 1930s; Cockcroft and Walton built a voltage multiplier machine at the Cavendish Laboratory, Cambridge, in 1931, and Lawrence and Livingston invented and built the first cyclic machine with a diameter of about 25 cm which accelerated protons to 1 MeV at Berkeley, California, in the same year.

It soon became clear that the way to higher particle energies was through the development of the cyclic machines. The machine of Lawrence and Livingston, which they called a cyclotron, operated on the principle of resonant acceleration. Although in its initial form (Fig. 7.1) the energy that could be reached was quite low, its mode of operation does serve to illustrate some of the principles of cyclic accelerators. The particles to be accelerated, for example protons, circulated in a vacuum between the poles of a large electromagnet and passed, twice in each revolution, across the gap between

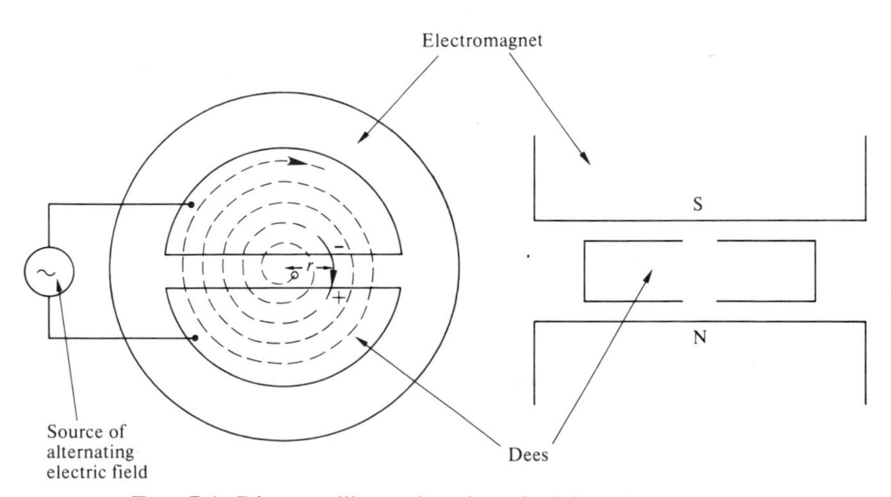

FIG. 7.1. Diagram illustrating the principles of a cyclotron.

two hollow, conducting electrodes called dees because of their shape. An alternating electric voltage applied to the dees in synchronism with the rotation of the protons delivered an accelerating electric impulse to the charged particles each time they crossed the gap. The purpose of the magnetic field was to maintain the particles in a trajectory which repeatedly crossed the accelerating electric field. (Inside the hollow metal dees there is no electric field.)

The motion is described by a simple formula (see note at end of chapter):

$$r = mv/Be,$$

where r is the radical position when the particle's speed is v; m is the particle mass, e its electric charge, and B the strength of the magnetic field. The time to make one revolution is:

$$2\pi m/Be,$$

which is independent of the particle's speed and the radius. Therefore, as the particle gains energy, the speed increases and its path spirals outwards, but the time between each crossing of the accelerating field is constant and the particle's motion remains in synchronism with an accelerating field of fixed frequency (10 to 20 MHz).

Protons can be accelerated up to an energy of about 20 MeV in this way but at this stage, when the proton's velocity has reached about one-fifth that of light, the synchronism condition begins to break down because of relativistic effects: no particle can exceed or even, if it has mass, reach the velocity of light. As the energy continues to increase, the relativistic mass rises but the velocity moves up more slowly and eventually becomes effectively constant, never quite reaching the velocity of light. Thus the time taken per revolution rises and the motion quickly falls out of step with the frequency of the accelerating voltage.

A way round this problem was soon found; the frequency could be slowed to keep in step. This resulted in the synchro-cyclotron which can take protons up to about 800 MeV, and such machines are still in use today. But now the limitation to reaching still higher energies became the size and cost of the electromagnet; so the next step was to abandon the idea of a spiral path in a massive magnet with a constant magnetic field and instead use an orbit of fixed radius with the confining magnetic field strength varying in time. This type of machine is called a synchrotron and the frequency of the accelerating system must follow exactly the rise of the magnetic field strength, so that the particles remain at constant radius throughout the acceleration process to a very high precision. This is achieved by an automatic feedback system which monitors the motion of the circulating bunches of particles and adjusts the accelerating field to hold the beam radius constant.

Nowadays the highest energy particle accelerators are all of the synchrotron type and consist of a large ring of magnets, dipoles, which bend the particles round a roughly circular orbit (Fig. 7.2). Other magnets, of a special type called quadrupoles, focus the particles, rather as a lens focuses rays of light, to form a narrow beam which circulates in a vacuum pipe placed in the jaws of the magnets. The accelerating fields are generated at radio frequencies and fed to resonant cavities placed in straight sections between arcs of bending magnets. When the particle's speed is close to that of light, c, the relationship between its energy, E, and the orbit radius becomes

$$E = ceBr.$$

So the maximum energy which can be reached is proportional to the fixed orbit radius and to the maximum magnetic field strength. The first is limited

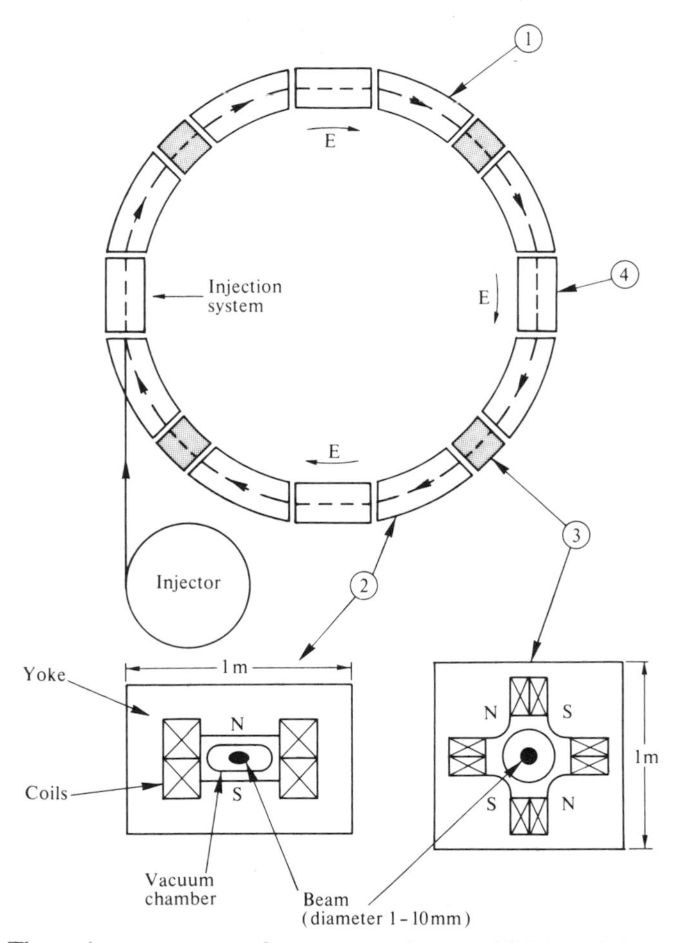

FIG. 7.2. The main components of a protonsynchroton: (1) Ring of electromagnets; as particles gain energy the magnetic field is increased to hold them in an orbit of fixed radius; (2) bending magnets (dipoles); (3) beam focusing magnets (quadrupoles); (4) RF cavities providing accelerating electric field.

by cost and the availability of land (although current practice is to tunnel unobtrusively below the surface) and the second, for conventional electromagnets with iron cores, by the value of B at which the iron becomes magnetically saturated. In practice the maximum energy which can be reached, with reasonable electric power consumption, by a large proton synchrotron is about 200 D GeV with D being the diameter in kilometres. The maximum energy could be taken higher by the use of electromagnets made with coils of superconducting wire operated at liquid helium temperatures. Magnetic field strengths about three times greater than is typical for conventional magnets can now be routinely obtained in this way, but so far the technical problems of manufacture and operation of the hundreds needed for a large machine with the necessary precision and reliability have not been completely solved.

There are two common types of machine. The first, called a fixed-target machine (Fig. 7.3), accelerates particles to a high energy and then they are ejected onto targets, usually narrow rods of metal. The accelerated particles

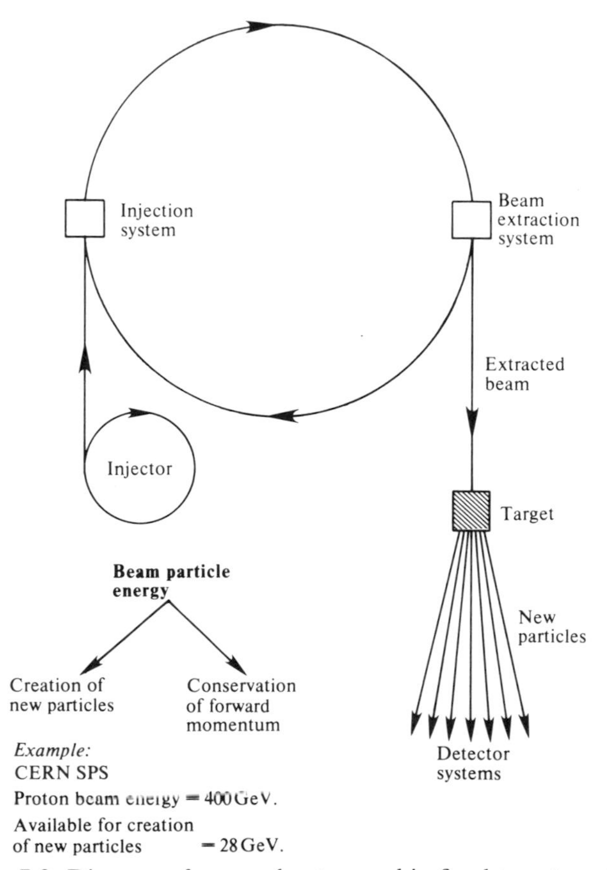

FIG. 7.3. Diagram of an accelerator used in fixed-target mode.

collide with the nuclei of atoms in the target and some of the energy they bring is transformed into the mass of many newly-created particles through the well known relationship $E = mc^2$. But, as the very large initial forward momentum of the incident beam particles has to be conserved, a lot of their energy has to be used to provide momentum for the products, mostly mesons, propelling them forward in a narrow cone, or jet. For example, when a proton of 400 GeV energy acclerated in the CERN SPS (Super Proton Synchrotron) collides with a proton 'at rest' only about 28 GeV is available for the creation of new matter. It is partly to get around this limitation that the second type of machine has been devised, in which the target is itself a beam of accelerated particles. In this device, called a colliding-beam machine (Fig. 7.4), the two beams are brought into head-on collision and, since the total initial momentum can now be zero (the two

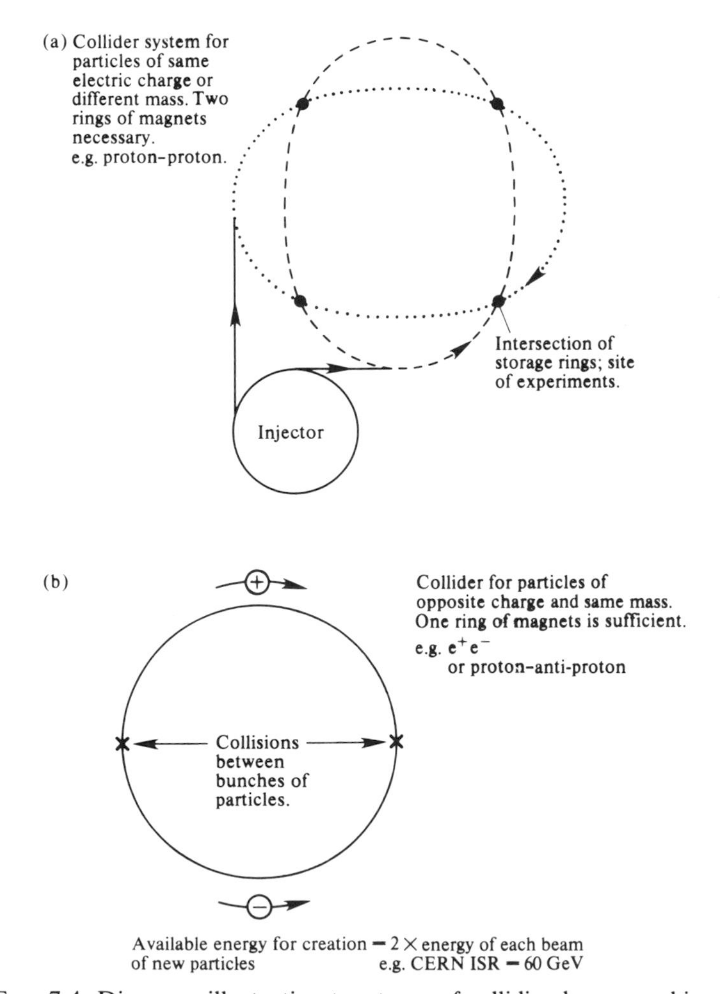

FIG. 7.4. Diagrams illustrating two types of colliding-beam machine.

incident momenta are equal but opposite), all the energy is in principle available for the creation of new particles. In the CERN ISR (Intersecting Storage Rings) two proton beams of maximum energy 32 GeV collide almost head-on so that the total available energy is nearly 64 GeV. A high available collision energy also means sensitivity to smaller-scale structure.

Four types of particle are stable enough to be accelerated and, in the case of colliding-beam machines, to be stored at high energy for many hours. These are protons and electrons and their antiparticles, the antiprotons and positrons. These six possibilities, two types of machine and four particles, can be combined in several ways to meet different needs of research. Proton fixed-target machines are the most common and are very versatile in the variety of secondary particle beams (of different mesons and leptons) which can be generated from the particles created in the primary target. Colliding-beam machines have so far been limited to electron–positron and proton–proton machines, but proton–antiproton collisions will be obtained at CERN in 1981 and plans are under discussion for electron–proton machines at several laboratories, including DESY at Hamburg.

At this stage, I would like to point out that these machines are not the 'atom smashers' of popular imagination. They do not so much break down atoms into small pieces as, rather, create new particles, even ones which can be much heavier than those colliding together. It is as though one banged two pocket watches together very violently and instead of finding only cog-wheels a grandfather clock emerged. Again, although we still call these machines accelerators, the particles in them are always moving very near to the velocity of light, which they cannot reach and, as we have already noted, their increase in energy is actually an increase in their relativistic-mass; thus one should perhaps call them 'ponderators' rather than 'accelerators', but the old name still sticks.

I can illustrate what such machines look like by using as examples the ones we have built at CERN. The two largest are the ISR, a proton–proton colliding-beam machine, and the SPS which is a fixed-target proton synchrotron. These two machines are fed with protons from a third, smaller proton synchrotron called the PS which was built first. It started operation in 1959, and itself consists of three machines operating in series: a linear accelerator which accelerates protons (from a gas discharge ion source) up to 50 MeV; a booster synchrotron which takes them on up to 800 MeV and a synchrotron which takes them up to a maximum energy of 28 GeV. The layout of this machine complex is shown in Fig. 7.5.

The colliding-beam machine, the ISR, consists of two rings of magnets which intersect at eight places. Each ring is filled with protons from the PS machine during a time of a few hours in order to build up the intensity of the beams of protons in each ring to a current of about 40 amperes (that is, about 8×10^{14} protons circulating in each beam). Thereafter, the two beams circulate round the machine for days on end colliding together at each of the eight intersection places. Very few protons actually meet on each revolution

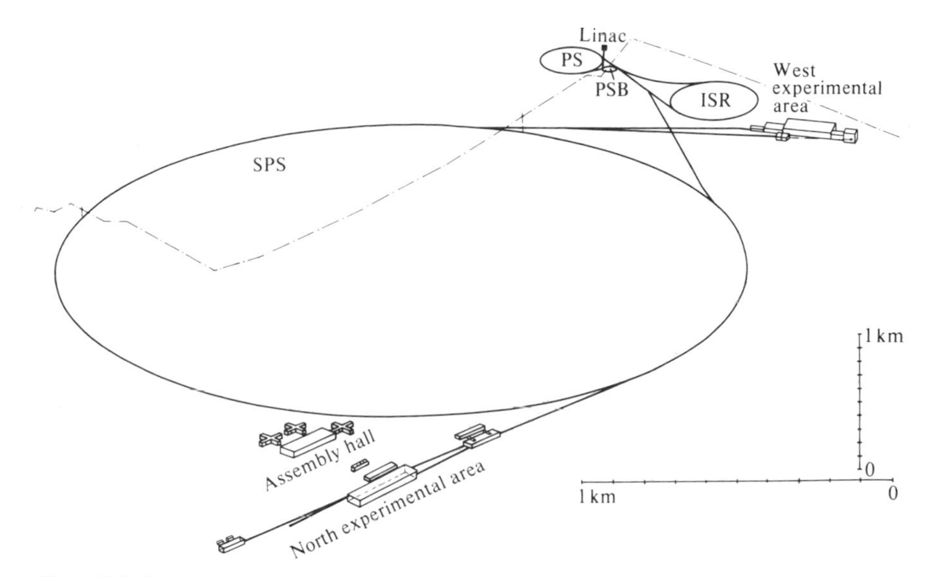

FIG. 7.5. Diagram showing layout of the CERN complex of PS, ISR, and SPS.

round the machine, so after a day most of them are still left in each beam. The main causes of loss of protons are beam instabilities, sending them out of orbit, and collisions with the residual gas in the vacuum chamber. To maximize the lifetime of the beams it is necessary to achieve a very low pressure in the vacuum chamber; pressures of 10^{-11} Torr are essential and this has led to new developments in ultra-high-vacuum technology. The experiments are carried out at the eight intersecting places where the beams collide and the particle detectors are built around these collision regions. The ISR machine is 300 m in diameter, can circulate protons up to 32 GeV energy in each beam and is the only machine of its type in the world, although, as I shall mention later, a much larger one is now under construction in the USA. Fig. 7.6 shows part of the two magnet rings of the ISR machine and a large particle detector installed at an intersection point; another detector and the ISR beam pipes are shown in Fig. 7.7.

The SPS is the biggest machine at CERN, with a diameter of 2.2 km, 7 km in circumference, and can accelerate protons up to 450 GeV. It was brought into operation for research in January 1977 and has been running 24 hours a day ever since with periodic stops for maintenance and modifications. There is only one other machine of this type and size in the world and that is at the Fermi National Laboratory, near Chicago.

The problems presented by the design and construction of the SPS are rather interesting from the technical point of view. In the first place, it is located about 50 m underground in a tunnel bored in the bedrock under the CERN site, and since the tunnel is only 4 m in section it had to follow the shape of the machine to within a tolerance of a few centimetres. Now boring

FIG. 7.6. The two rings of the ISR converge at one of the eight intersections where a large particle detector (the SFM) is placed. (CERN).

tunnels to this accuracy is not commonplace. Engineers who make railway or road tunnels from either side of a mountain are usually content if they join up to within a metre or two, but we had to ensure that our tunnel was in the right place all the way round its 7 km circumference to within a few centimetres. The tunnelling machine we used was a full-face rotating-head boring machine weighing about 200 tons which could tunnel at a rate of 20 m a day. Fig. 7.8 is a photograph of this machine.

The problem was how to guide this monster between the eight places about 1.2 km apart, where we had vertical shafts coming to the surface from which we could check its position from a ground-level survey grid. The solution adopted was to use the earth's rotational axis as the reference and to guide the machine with a gyrotheodolite and laser beams. The gyrotheodolite took its bearing from the earth's rotational axis and was used to set the direction of the laser beam, which in turn was used to orient the boring machine metre by metre as it chewed its way round the tunnel on a curved course. The system worked very well and the tunnelling machine arrived back at its starting point within a few centimetres' accuracy after boring 7 km of tunnel in 15 months.

The electromagnets which bend the protons round the machine and focus the circulating beam of particles through the acceleration process are

FIG. 7.7. The ISR vacuum pipes. One carries the beam entering, the other leaving an intersection surrounded by a particle detector. (CERN).

arranged in a periodic sequence which is called the machine lattice. Each period of the lattice consists of four bending magnets, a focusing magnet or quadrupole, another four bending magnets and then another quadrupole, and there are 108 of these lattice periods around the circumference of the machine. The quadrupoles focus the beam alternately in the vertical and horizontal planes. In all, there are therefore 216 quadrupoles and over 744 bending magnets. The problem was how to manufacture all these magnets to a very high precision. The tolerable variation, for example, in the magnetic field from one bending magnet to another is about one part in ten thousand and this tolerance must be kept as the magnetic field rises from 450 to 18 000 gauss. The magnets must also be aligned very precisely in the tunnel, and for the focusing magnets, for example, a positional tolerance of one tenth of a millimetre is required. Such accuracy naturally calls for a very precise way of measuring the position of the machine in the tunnel. This was achieved by

FIG. 7.8. A photograph of the boring machine used to make the 7 km ring tunnel for the SPS. (CERN.)

fixing reference points on the walls of the tunnel all the way round its circumference, measuring the distance between them very precisely with invar wires and feeding all the data into a very large computer which calculated the position of each reference point relative to the ideal machine axis. Using these reference points we could then adjust the position of all the magnets. When we circulated a beam of protons round the machine later on we found that we had indeed achieved the accuracy we required. It is interesting to point out that the SPS machine is so large that allowance has to be made for the curvature of the earth in levelling the bending magnets of the machine. Each magnet had to be tilted slightly with reference to the local vertical to get all the magnets in the same plane. Fig. 7.9 shows the components of the SPS machine installed in the underground tunnel. During their acceleration, which takes nearly two seconds, the protons travel more than 500 000 km, a distance greater than that to the Moon, yet during this long passage they do not deviate from the central orbit of the machine by more than a millimetre.

The accelerating system of the SPS machine works at a frequency of 200 MHz, since there are about 4600 proton bunches rotating round the machine and the revolution time is 23 microseconds. To create the necessary accelerating voltage of 5 million volts per revolution requires a radio-

FIG. 7.9. CERN SPS magnets in their underground tunnel. (CERN.)

frequency power of 2 MW. One of the two accelerating structures, called an RF cavity, is shown in Fig. 7.10.

To operate this giant accelerator whose components are distributed around its circumference we use a computer control system enabling just one or two operators to control the whole machine. Information is collected from the machine components and control instructions are sent out to them via a series of about 30 small computers which communicate with each other

FIG. 7.10. One of the two radio-frequency accelerating structures for the CERN SPS. (CERN).

by a sort of high-speed telephone exchange called a message switching system. The operator receives data on television-type screens and can send instructions to any component of the machine by means of a touch panel, which identifies the component, and one knob, which determines the required action. It is a system which was developed at CERN using commercially available computers and it makes the job of operating the SPS particularly straightforward.

The total cost of the SPS machine was about £250 million. The management of such a large construction project clearly calls for modern methods of management control, Many thousands of firms in Europe were involved in the manufacture of the components of the SPS machine and in its installation. The design of the SPS was carried out by the CERN project team who were also responsible for supervising the manufacture of its components in industry, for their installation at CERN, and for the commissioning of the sub-systems of the SPS and the complete accelerator. The same team is now operating and developing the machine. It was completed in six years, somewhat quicker than the time schedule laid down for its construction, and within the original financial provisions; something which is not so common with very large engineering projects involving advanced technology.

It is important to point out the differences between these two machines, thc ISR and the SPS. Although, as already noted, the available energy in collisions produced in targets by the 400 GeV protons of the SPS is only about half that attained in the ISR at their maximum energy, the number of collisions occurring per second in the SPS target is about 10 000 times greater than is achieved when the ISR beams pass through each other. This higher rate of collisions in the SPS is very important for certain rare processes and together with the greater variety of interactions that can be studied is a compensating advantage; from the research point of view, they are complementary.

We are now in the course of modifying the SPS to accelerate and store antiprotons. Since antiprotons have the same mass as protons but the opposite electric charge, both particles can be made to circulate in the same vacuum chamber of the single ring of magnets of the SPS at the same time, but in opposite directions. In fact, protons will circulate clockwise and antiprotons anticlockwise. Thus, starting in 1981, the SPS machine will also be used as a proton–antiproton colliding-beam machine and produce collisions in which the available reaction energy is as high as 600 GeV, about ten times as much as that possible in the ISR machine.

So far the examples chosen to illustrate particle accelerators have all been proton machines but another very important class are those in which the particles are electrons. The acceleration of electrons, or positrons, poses a unique problem which is a consequence of their very small mass. When bent round a circular path by a magnetic field all charged particles lose energy in the form of synchrotron radiation and this lost energy has to be more than replaced by the accelerating system. Protons, even at energies as high as 400 GeV, radiate very little energy in this way and it is a negligible problem for the proton accelerators. Electrons and positrons however are nearly 2000 times lighter than protons and, at the same energy, radiate a million million times more; the amount radiated increases as the fourth power of the particle energy. This imposes a very serious limitation on the maximum energy which can be reached in a circular machine. One solution is to use a linear accelerator in which the electrons pass, once, through a chain of accelerating cavities. This is the case for the Stanford Linear Accelerator (SLAC) in California which is two miles long; it was completed in 1966 to accelerate electrons to 20 GeV; improvements to the RF power supplies should allow about 45 GeV to be reached.

Storage rings for the study of electron–positron collisions have been responsible for several important discoveries in recent years. At present the highest energy, about 19.5 GeV per beam, has been reached in the PETRA machine at the DESY Laboratory, Hamburg; the recently commissioned machine, called PEP, at Stanford, will reach similar energies.

Particle detectors

I would like now to pass on from the particle accelerators to the particle detectors and very briefly give some idea of what they are like. Remember that a detector has to measure all the products of a collision between two particles or as many of them as is possible. One can divide them into two types: visual and electronic. With visual detectors the trajectories of charged particles participating in the collision are made visible to the eye. Bubble chambers are typical of this type; in these devices the tracks of charged particles appear as lines of very small bubbles created in a liquid close to its boiling point. Commonly used liquids are liquid hydrogen and liquid propane, both of them highly explosive and demanding the strictest safety precautions. In fact one does not just observe collisions in a bubble chamber, but one takes photographs of them and the output of such an experiment consists of millions of photographs which have to be scanned and then measured in order to reconstruct in three-dimensional space what happened in the collisions. A bubble chamber operates in a strong magnetic field which bends the trajectories into arcs of circles, enabling the momentum, energy, and charge of the particles involved in the collision to be measured and so the different types of particles to be identified. Fig. 4.6 shows a general view of the largest bubble chamber operating at CERN and Fig. 4.8 is a photograph of tracks of particles in this detector.

In electronic detectors, on the other hand, the particle trajectories are not usually visible but can be reconstructed from electronic signals. A large number of different devices have been developed which can indicate electronically the passage of a particle and its geometric position in space. By combining these devices in different ways with magnetic fields one can reconstruct in space the trajectories of particles taking part in the collisions, measure their momenta and electrical charge, and identify them. The output of electronic detectors is millions of electronic signals which can be stored on magnetic tapes in digital form to be subsequently analysed by a computer system.

As an example of an electronic detector which has a very important place in most experiments today we can take the multiwire drift-chamber. The principle is illustrated in Fig. 7.11 which is a schematic cross-sectional view of a basic 'cell' which can be repeated many times to cover a large plane. As a charged particle passes through the chamber it knocks electrons out of the gas atoms; these are attracted to the nearest anode wire and on arrival an electrical impulse is caused which can be detected and recorded by the electronic data-collection system. The electrons drift towards the anode at a known speed, about 1 mm in 20 nanoseconds (20×10^{-9} s), and so if the time of the particle's arrival is also measured the drift time gives the distance of the trajectory from the anode wire. The arrival time can be determined by a signal from another type of detector, a scintillation counter, placed in front of the multiwire drift chamber. The scintillation counter is the most common electronic detector; the disturbance of the atoms by a passing charged particle causes flashes of light to be emitted, scintillations, and these are picked up, with negligible delay, by photomultipliers. As described, one multiwire plane only measures one coordinate accurately, the one perpendicular to the anode wires; by using a sequence of planes with wires at different orientations accurate spatial coordinates can be obtained without

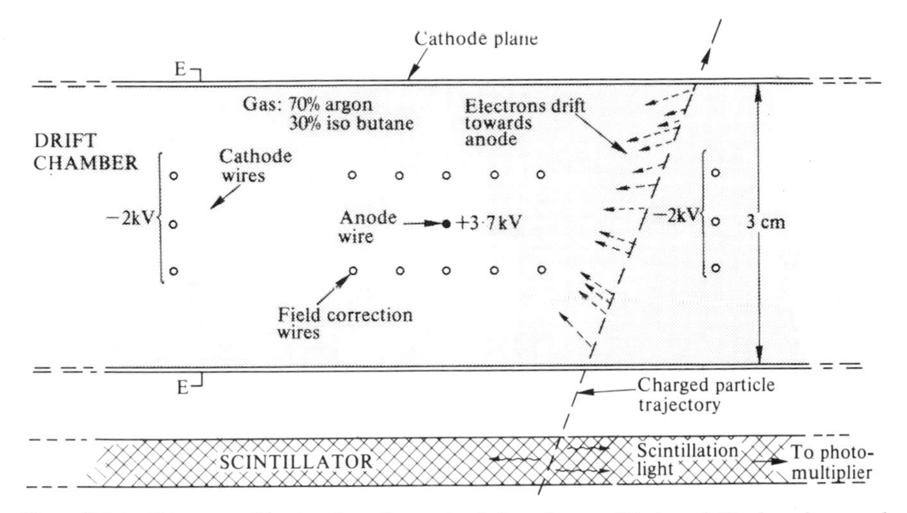

FIG. 7.11. Diagram illustrating the principle of a multiwire drift-chamber and scintillation counter. The drawing shows a cross section of one 'cell' of a chamber similar to those used in the neutrino experiment pictured in Fig. 4.7.

ambiguity. Accuracies of about 0.1 mm can be obtained and some of the bigger chambers cover areas of 5 m by 5 m. For a single experiment tens of thousands of wires, many only 0.025 mm in diameter, may have to be strung on large frames with a precision of about 0.050 mm.

To identify a particle it is necessary to measure its mass and this can be obtained from its momentum if its velocity is measured (momentum depends on mass and velocity only). In all these experiments most of the particles of interest are moving at velocities near that of light, and the most successful method of velocity measurement is based on detection of Cerenkov radiation. This is light emitted by a fast-moving charged particle passing through a transparent medium when its velocity is greater than a threshold value. (The threshold is actually the velocity of light divided by the refractive index of the medium and in a way the Cerenkov light is analogous to the sonic boom radiated by an aircraft flying faster than the speed of sound in air.) The light so emitted is picked up by photomultipliers and transformed into electronic signals. Cerenkov detectors for very high-energy particles can be very long, 10 m or more.

The array of electronic detectors used in an experiment can generate a huge amount of data. For example, the digital signals from one collision may be of the order of 10 000 bits and the number of collisions may be as many as one million a second. One advantage of electronic detectors over bubble chamber detectors is that a selection can be made of the electronic signals that are processed. If one knows what one is looking for, one can decide in advance which electronic signals are worth recording and analysing, and very large reductions in the amount of data can be made. This data selection, however, has to be done in stages and at high speed and this has led to the use of on-line computers directly attached to the experiment. More recently microprocessors have become available which greatly facilitate this local selection and processing of data, but, so far at least, one still needs a powerful central computer system for the treatment of the data from experiments using electronic detectors. The electronic selection of data has its disadvantages; it may introduce biases in the experiment by unintentionally excluding relevant data or, even worse, it can discard interesting phenomena taking place in the collisions. In this respect, the bubble chamber, by photographing everything on film, at least preserves all the information and the selection is applied non-destructively when analysing the film.

Whereas the useful volume of bubble chambers is limited in size by engineering considerations—the largest at CERN (BEBC, Fig. 46) measures 3.7 m in diameter—the size of an electronic detector system is in principle unlimited since its various components can be spread out in space and more added as required. This is particularly important at the very high collision energies that are being studied nowadays, in which the particle trajectories must be measured over distances of tens of metres. Another important factor is mass. Neutrinos have such a weak interaction with matter that intense beams of them may pass through hundreds of tons of material without anything happening. Bubble chamber experiments with neutrinos have been very successful and for this purpose BEBC is often filled with 10 tons of a liquid neon–hydrogen mixture. But large numbers of neutrino interactions can only be accumulated, and so rare processes studied, with an electronic detector system like that shown in Fig. 4.7. This consists of a multi-layer sandwich of magnetized iron discs (forming a torroidal magnetic field) interleaved with large multiwire drift chambers and planes of scintillation counters. Its total length is about 30 m and it weighs over 1500 tons.

A recent development is the use of bubble chambers to photograph the region close to the point of collision, or vertex, where the particles are close together, and to use electronic detectors to measure the ongoing trajectories outside the bubble chamber. In this way one retains the precision of the bubble chamber to observe the fine detail of the collision vertex while extending the measurement and identification of the particles for tens of metres along their subsequent path.

In the case of a colliding-beam machine, as I have mentioned, the detectors are built around the places in the machine where the two beams intersect. In a fixed-target accelerator, on the other hand, between the place at which the proton beam is extracted from the machine and the experiment there is a distance of one or two kilometres which is taken up in preparing the particle beams required by each experiment. In the case of the SPS machine, which is some 50 metres underground, the extracted proton beam from the machine has to be brought to the surface where the experiments are located. The SPS feeds protons to about nine targets which in turn produce different kinds of particle beams for about 20 experiments. On each accelerating cycle of the machine, that is to say every 10 seconds, each target is fed with protons of the required energy and intensity in a very complicated operation which, like the accelerator, is controlled by the SPS computer control system. In chapter 4 Donald Perkins has described the way in which a pure neutrino beam is produced. The distance required, from the target placed in the extracted proton beam, is about 1 km. Incidentally, since neutrinos are not electrically charged, and cannot therefore be guided by electric or magnetic fields, the proton beam from which they are generated has to be pointed towards the particle detectors very carefully so that the neutrino beam goes through the centre of the bubble chamber and the three electronic detectors which are placed one behind the other along the neutrino beam.

Data analysis

The third of the tools of the experimental physicist is the data analysis system and about this I will be very brief. It should be clear by now that the detectors used in experiments around the ISR and SPS machines at CERN produce vast amounts of data; millions of photographs from bubble chambers and even more millions of digital signals from the electronic detectors. Furthermore, the rate at which this data is produced is very high. Most of this data can be stored either on photographic film or on magnetic tapes (one experiment may fill several thousand tapes in a year), but it all has to be analysed if results are to be obtained from the experiments. In addition, some of the data has to be processed as it is produced in order to select events for final analysis and to check that the experiment is running correctly. Clearly, the data analysis system must use very large and very fast computers and at CERN there are two of these computer systems. But since the CERN experimenters come from the European universities and as academic staff have also to teach, there is every advantage in sending the data from CERN to the universities and analysing it there. In fact, only about one-third of the data produced at CERN is analysed on site. However, universities cannot usually afford the big computers needed, so national centres have been set up in Europe for this purpose. These are at least closer to the universities than CERN, and the universities are often connected by computer links to them. We are even experimenting now at CERN with sending out experimental data at very high speeds via the European communications satellite to national centres, such as the Rutherford Laboratory, each centre being equipped with small ground stations for this purpose. This experiment is supported by the European Economic Community as a test-bed for the European communications industry.

Finally, it is important to point out that the three main tools of the experimental physicists, the accelerator, the detector systems, and the data analysis system, must be continuously optimized in their operation so that the flow of information is not impeded. The accelerators must produce enough protons for all the experiments, and the computers, wherever they are located, must be capable of processing all the data in a reasonable time. It is a continuous struggle to keep the whole system in balance without bottlenecks occurring.

Problems of size

I would now like to break away from the technology of experimenting at very high particle energies and mention some of the problems which arise due to carrying out experimental physics on this scale.

The first point to be made is that it is the research itself which has determined the present scale of operations and one is sure that if there were any way of carrying on this research without the giant machines, the massive detectors, and the big computers, the experimenters would prefer it. But unfortunately there is no other known way; the need for high energies to explore the small-scale structure of matter is a fundamental principle of physics and we know of no technology which would permit accelerators to be much smaller or cheaper. The experimental equipment must also be large and make use of the latest advances in electronic technology. So one has to learn how to keep the spirit of pure research alive and healthy amidst all this machinery.

I have already mentioned that the size of the machines has led first to National Laboratories and then to International Laboratores. In Europe, most of the National accelerators for high-energy particle physics have now been closed down and apart from the European Laboratory at CERN only one National Laboratory, DESY, at Hamburg, is now operating a front-line machine, an electron–positron colliding-beam machine. This concentration of the research machinery in one or two places in Europe produces many problems.

The first is the way in which experimenters carry out their research. In many fields of scientific endeavour a few scientists can conceive an experiment, put the apparatus together in a few months, carry out the experiments on a similar timescale, and publish the result shortly after. If the experiment fails in any way it is only a local disappointment and one can quickly pass on to another. It was like this in nuclear particle physics up to the Second World War. Nowadays, an experimenter is usually a member of a team of twenty to fifty or more physicists coming from several universities in different countries. The detectors he uses are often built by the team with a considerable contribution from professional and technical staff in the universities and central laboratories like the Rutherford Laboratory. It takes three to six years to construct, install and carry out an experiment and to analyse the data. Publication of the results is also a lengthy business and the first page of the publication is often taken up with the list of authors and their university affiliations. Failure of an experiment to produce valid results affects many people for several years and is public knowledge in the world community of physicists. Clearly, this style of research raises the problem of the place of the individual physicist in this research. One might imagine that a young physicist would lose his identity as a creative scientist in these large teams, but this does not seem to be the case. Despite having to spend so much time on technology the creativity of experimenters seems to be as high as ever, new ideas are always coming forward at a fast rate and brilliant young physicists are still entering this field of research in much the same numbers as ever. Partly this is due to the high intellectual challenge and very fundamental nature of the research, partly it seems to be that young physicists adapt very quickly to all the technology involved and not only enjoy it but feel that it gives them an excellent training should they leave research for industrial employment later on, which many of them do. In any case, one hears little complaint in this respect; there is much more worry these days about finding more stable employment in universities for those physicists who want to make their career in experimental high-energy particle physics.

A second problem concerns the organization which is necessary in order to carry out research on this scale. A Laboratory like CERN employs about 3500 staff, of which less than 100 are research physicists. The physicists performing experiments at CERN come from the European universities and number about 1500. The Laboratory's annual budget is about 600 MSF. Clearly, this is an enterprise requiring considerable organization and the problem is to maintain an atmosphere in the Laboratory conducive to creative research while at the same time providing an economic and efficient management to handle all the technological operations and to control the financial expenditures. When one remembers that on top of this CERN is an international organization with twelve Member States each with sovereign powers, it is remarkable that one can carry out fundamental research at all in this way and even more surprising that it advances so quickly.

The third problem is how to select experiments to be carried out in the central laboratory. The system which has evolved is the following. Proposals for experiments are put forward by the teams of physicists which, as I have explained, are composed of university physicists from many universities and several countries. The proposals are first discussed at open seminars at CERN where the team presents its proposal to a highly critical audience of fellow physicists. The idea may not survive this ordeal or it may be modified and improved. If it survives, it then goes to what is called an Experiments Committee, of which we have one for each major accelerator at CERN, and there is further criticism and suggestion. Only then does it reach the Research Board and possibly final approval, when it is allocated time in the operation schedule of the machine. At this stage the construction of the experimental apparatus can begin and it may start running a year or two later. This method of selecting experiments is highly democratic, a sort of successive judgment by peers, and one wonders how many of the brilliant experimenters of the past, Rutherford for example, would have reacted if they had had to suffer such an examination before they were allowed to start an experiment. The method also raises doubts about whether the experiments so selected are the best. Does the system favour good experiments which will certainly yield valuable results and disfavour risky experiments whose results might be of future Nobel Prize quality? Is a dynamic and articulate team leader more likely to persuade his peers than a shy, reticent leader who may nevertheless be proposing a better experiment? We have no evidence that any such biases exist in the system once a proposal is submitted, but one never knows whether the system determines to some extent the proposals which are submitted.

My last example is the problem of how to carry out experiments at CERN and teach in a university. It is generally accepted that fundamental research is an essential part of university activities with a vital impact on the quality and content of university education. Experimenters in high-energy physics, like most other university staff, teach as well as carry out research. However, unlike most others, they have to do their experiments far from their university and in another country. Furthermore, the experiments cannot just be fitted in during the long vacation periods, they run for several years and the team works twenty-four hours a day in shifts for months on end. It is difficult to combine teaching and research on this basis, and now that university posts are in such short supply all over Europe it becomes more and more difficult to maintain an effective presence at CERN and at the university throughout the academic year. Nevertheless, there is no lack of experimental physicists wanting to use the CERN facilities and very few people would contemplate any separation between teaching and research in this fundamental field. I can only conclude that the fascination of the research more than compensates for the hardships involved.

Future tools

I would like to conclude with a few remarks about the future. The position at the present time is as follows. There are three regions of the world which are most active in this field of research, the United States of America, the Soviet Union, and Western Europe. Up to now there has been some duplication of the front-line machines, although rather little in the last ten years, but in the future the machines needed are so large that one looks for complementarity rather than duplication.

Two projects now under way should be completed within the first half of the 1980s. One is the modification of the SPS at CERN so that it can be used as a proton–antiproton collider producing collisions with total available energy up to about 600 GeV; this is due to start operation in 1981. Meanwhile, at the Fermi National Laboratory in the United States, a ring of superconducting magnets is to be added next to their existing (500 GeV) proton synchrotron. This will enable protons to be accelerated to 1000 GeV

(1 TeV) for fixed-target experiments, or it could be used as a proton–antiproton collider of total energy up to 2 TeV.

Looking beyond, to the second half of the 1980s, the USA is constructing a very large proton–proton colliding-beam machine at Brookhaven on Long Island, called ISABELLE, also using superconducting magnets, which aims at energies of 400 GeV in each beam, or 800 GeV collision energy, with very high beam intensities. The Soviet Union has just decided to build a very large fixed-target proton machine, called UNK, 20 km in circumference, to give beam energies up to 3000 GeV. The two latter machines are similar to, but much larger than, the ISR and SPS at CERN. European physicists want to build a very large electron–positron colliding beam machine, called LEP, 30 km in circumference and capable of the acceleration of electrons and positrons to about 130 GeV in its fully equipped final phase. A project for such a machine has been designed at CERN to be built next to the existing laboratory. Fig. 7.12 shows the layout of this machine on the CERN site. We hope to get approval for this project in 1981 and to complete it during the second half of the 1980s. It is interesting to note that the PS/SPS machine complex will be modified to accelerate electrons and positrons so that it will also act as the injector for the future LEP machine. In West Germany a proposal to build a high-energy electron–proton collider, called HERA, at the DESY Laboratory, Hamburg, is under discussion.

From the research point of view, the purpose of LEP is to study what happens in the energy range where the electromagnetic and weak nuclear interactions become comparable in strength. The new unified theory of these two interactions, discussed in chapter 5 by Abdus Salam, predicts the existence of new particles called intermediate vector bosons with masses of about 80 GeV and 90 GeV. These particles could be produced by LEP in numbers which would enable their study under very favourable experimental conditions. The SPS proton–antiproton collider, mentioned previously, may be able to discover whether these bosons exist but the LEP machine is needed to explore in detail the complete physical behaviour of the weak and electromagnetic interactions in this very important energy region.

The LEP machine presents many interesting technical problems, some of them due to its immense size and the need to minimize the cost of its components to keep its total cost as low as possible. But the major problem for LEP, as with all circular electron machines, is the synchrotron radiation energy loss we have already discussed. The large size of LEP is due not to its particle energy but to the need to reduce synchrotron radiation losses and to economize on electrical power consumption. The energy radiated by a circulating particle increases rapidly with its energy but is inversely proportional to the radius of curvature of its path in the machine; the larger the machine radius the less the electrons and positrons radiate energy. This is why the LEP machine, designed for maximum electron energies of about

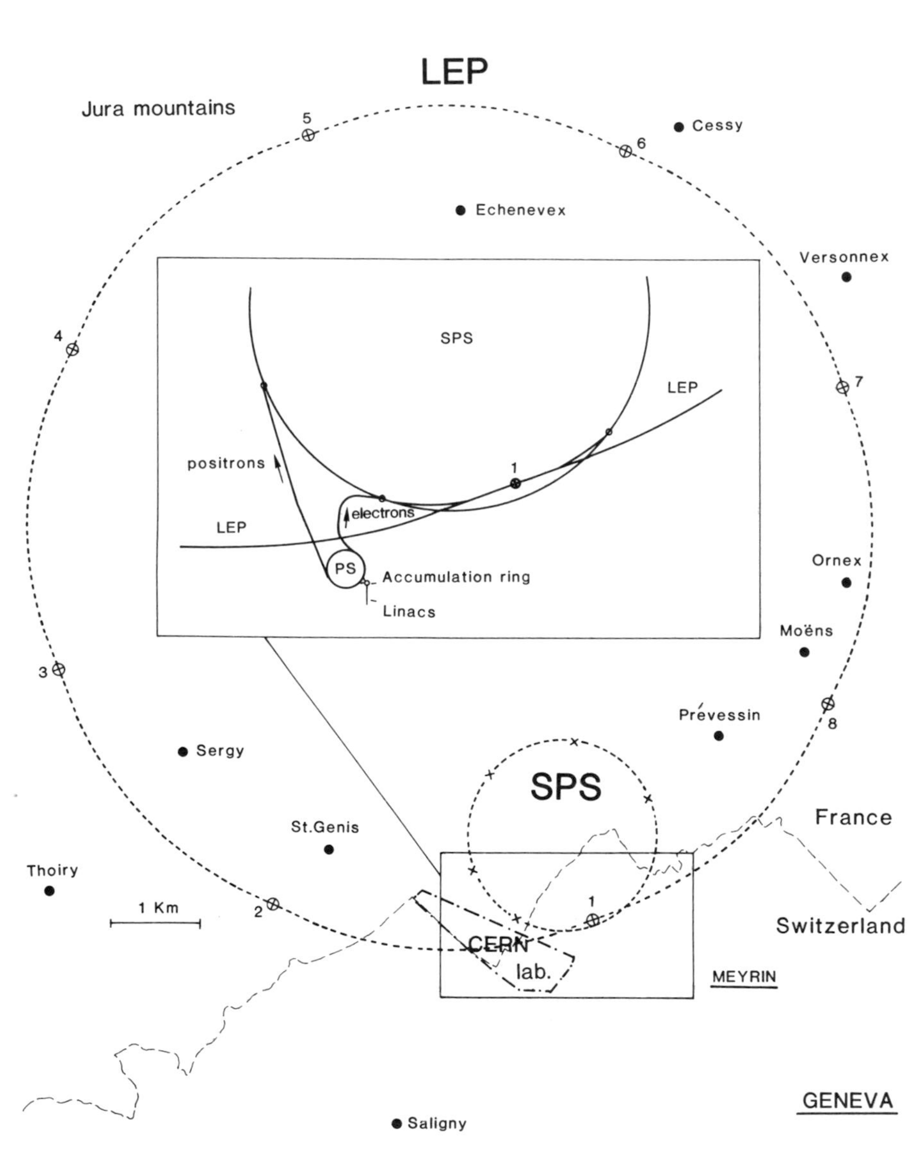

FIG. 7.12. The site of the proposed electron–positron collider LEP. Its tunnel, 30 km in circumference, would reach under the Jura mountains just west of Geneva and the existing CERN PS and SPS would be used as injectors. The ring is not an exact circle; there are long straight sections where the beams pass through the eight intersection points.

130 GeV, is 30 km in circumference, whereas the SPS machine is four times smaller but can accelerate protons to 450 GeV. Even so, at an energy of 82 GeV per beam the power input of the LEP accelerating system is about 150 MW. Such a high power consumption is clearly unwelcome at the present time when energy resources are becoming a serious problem in Europe; consequently, the design of the LEP machine incorporates many new ideas to minimize its electric power consumption: the most advanced of these is the development of superconducting radio-frequency cavities for the accelerating system to reduce ohmic losses in these cavities.

One might ask, why not avoid synchrotron radiation by building a linear accelerator, as at SLAC? But such a system cannot store the electrons and positrons and although the idea of firing the beams of two very long linear accelerators at each other is being studied it does not appear economically feasible unless it is decided that much higher energies than LEP become necessary.

The financing of a huge machine such as LEP also raises serious problems. In the present economic situation in Europe it is not possible to imagine that the Member States of CERN will be willing to increase the CERN budgets substantially. Consequently, we must plan to build it within the present annual budget levels of CERN and this means closing down other machines, such as the ISR, in order to find the money and manpower to build LEP.

When these new machines, the Fermilab Tevatron/Collider, ISABELLE, UNK, LEP and perhaps HERA are all in operation towards the end of the 1980s we will have in the world four or five laboratories with major front-line machines which together will cover the needs of the research. But since they are different types of machine experimenters will have to travel not only within their own region but also to other regions, depending on the experiments they wish to do.

Even today there is a considerable amount of sharing of research facilities between the three regions. At CERN one finds not only European university physicists but also Soviet and American physicists, and European physicists are to be found in America and in the Soviet Union. In the future this sharing will increase and perhaps one day, towards the end of this century, if the machines get bigger still and depending on the evolution of international politics, one may find only one World Laboratory being used by all the physicists carrying out research into the fundamental constituents of matter.

Note on the motion of charged particles in a magnetic field

A particle of charge e travelling with speed v perpendicular to the lines of force of a magnetic field of strength B experiences a force of magnitude:

$$\text{force} = Bev.$$

The direction of the force is always perpendicular both to the field lines and to the instantaneous direction of motion of the particle. This latter property of the force is just what is required to cause a particle to move in a circle and the laws of elementary mechanics tell us that to maintain motion in a circle the magnitude of such a force must be:

$$mv^2/r,$$

where m is the particle mass and r the radius of the circle.

The equation describing the motion is therefore

$$mv^2r = Bev,$$

or

$$r = mv/Be = \text{momentum}/Be.$$

8

Questions for the Future

MURRAY GELL-MANN

In this review of present ideas and questions for the future my theme will be the search for unity in our description of elementary particle physics. Everything in the Universe, including ourselves, is made of elementary particles, each kind behaving in exactly the same way in every part of the Universe, as far as we can tell from the light reaching us from the most distant galaxies. The laws of these elementary particles and of the forces governing their behaviour constitute the fundamental principles of microscopic physics. If we adjoin to these the principles of cosmology, in particular the boundary condition that about 15 000 million years ago the Universe was a tiny, hot, dense, expanding ball, then we have all the fundamental laws of physics. They underlie not only the rest of physics, but also astronomy, chemistry, geology, biology, . . . in fact all of natural science. When Dirac wrote his relativistic equation for the electron in 1928 he commented modestly that his equation explained most of physics and the whole of chemistry.

In this enterprise of discovery, as in natural sciences generally, the experimental and theoretical contributions go hand in hand; these days they are usually made by different sets of people. Sometimes the theorists are ahead, with correct predictions, but occasionally the experimentalists spring a surprise that sends the theorists back to the drawing board. I do not have time to discuss the experimental data supporting the theoretical ideas that I will mention; some of them have been described in earlier chapters. I will, however, take care to distinguish those ideas that have some appreciable experimental support and are probably right, or nearly right, from those ideas that are highly speculative and must be tested in future experiments. It is very important to draw that distinction because the two kinds of ideas may sound equally crazy.

In elementary particle theory one assumes the validity of three principles that appear to be exactly correct.

(1) *Quantum mechanics*, that mysterious, confusing discipline, which

none of us really understands but which we know how to use. It works perfectly, as far as we can tell, in describing physical reality, but it is a 'counter-intuitive discipline', as the social scientists would say. Quantum mechanics is not a theory, but rather a framework within which we believe any correct theory must fit.

(2) *Relativity*. Seventy-five years after Einstein's first work on relativity, we have no reason to doubt it.

(3) *Causality*, the simple principle that causes must precede their effects.

These three principles together constitute the basis of *Quantum Field Theory* and all respectable speculation in our field is carried out in the context of quantum field theories. Here each force is communicated through the exchange of a *quantum*; for example, for electromagnetism that quantum is the photon. The quantum field theory of electrons and photons has existed for a little over 50 years and is called quantum electrodynamics. It deserves its nickname of QED because it agrees fantastically well with experiments, to a precision of many decimal places. In QED, as in other quantum field theories, we can use the little pictures invented by my colleague Richard Feynman, which are supposed to give the illusion of understanding what is going on in quantum field theory. In Fig. 8.1, for example, an electron emits a photon which is then absorbed by another electron, giving rise to the electromagnetic force between the two electrons. Strict energy and momentum conservation would appear to forbid the one electron from emitting the photon and the other from absorbing it; however, in the Pickwickian sense of quantum mechanics such happenings are allowed and the effect is a force between the two electrons.

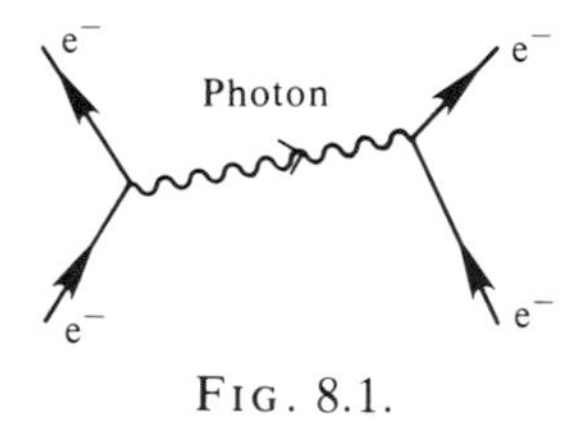

FIG. 8.1.

Matter as we see it is made of molecules or crystals, and these are made of atoms or ions, which in turn are made of nuclei surrounded by electrons. The nuclei, as we have known for almost 50 years, are made of neutrons and protons. All of these are composite structures (except the electrons) and the proton and neutron are made, in their turn, of quarks. (Quark is an obvious name for the fundamental constituents of the neutron and proton!) The recipe for making a neutron out of quarks is to take one of charge +2/3, called a u quark, and two each of charge −1/3, d quarks; for the proton just interchange u and d quarks. The total charges add to 0 and +1 respectively. (We use units in which the proton charge is +1; the sign was determined arbitrarily by Benjamin Franklin.)

Table 8.1

Forces in quantum field theory

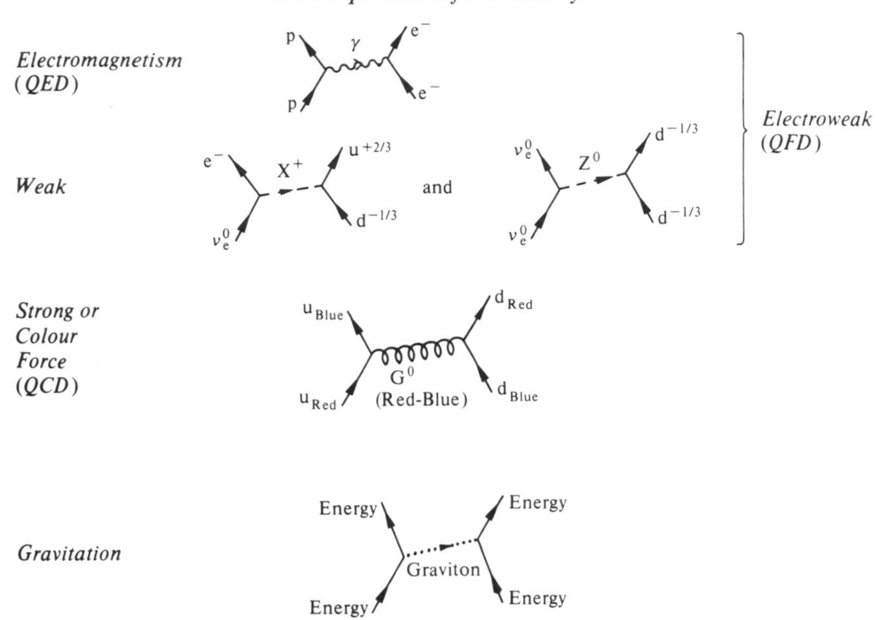

The u and d quarks are known as 'flavours' of quark and there are other flavours, in addition to these two.

There is another distinguishing characteristic of quarks, known by another pet name, 'colour'. Colour has no more to do with real colour than flavour has to do with real flavour; they are just convenient names for physical variables each with several values. In the case of colour we know how many values: exactly three. So we can name the three colours arbitrarily 'red', 'green', and 'blue', which are supposed to be the primary colours of human vision.

To continue the recipe for building protons and neutrons, the three quarks we take must each have a different colour, and they are assembled in

Table 8.2

Neutron and proton are made of quarks			
Neutron n^0	$u^{+2/3}$	$d^{-1/3}$	$d^{-1/3}$
Proton p^{+1}	$d^{-1/3}$	$u^{+2/3}$	$u^{+2/3}$

There is one quark in each of three 'colour' states (say: 'red', 'green', or 'blue') and the over-all colour of the proton or neutron must always be neutral (or 'white').

such a way that the colour 'averages out'; the resulting neutron or proton must have no net colour. Continuing the metaphor, it must be 'white'.

The quarks naturally have to be held together by some kind of force, and, in quantum field theory, that force has to arise from the exchange of certain quanta. In this case they are called 'gluons', because they glue the quarks together inside the neutron and proton. Again we can draw a Feynman diagram (Fig. 8.2), showing the exchange of a gluon between quarks.

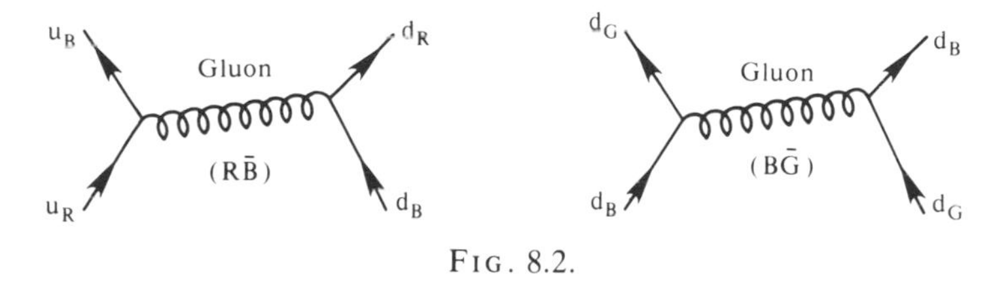

FIG. 8.2.

The flavour makes no difference because the gluon forces are completely indifferent to flavour. However colour is very important; for different colour situations we have different gluons. There are eight different sorts of gluon corresponding to different colour combinations.

A definite quantum field theory has been proposed over the past ten or fifteen years for the coupling of quarks and gluons; it is called quantum colour dynamics or quantum chromodynamics (QCD) and is closely analogous to QED. So far it seems to be working extremely well and it is probably right. The analogy can be seen in the following way: instead of the electron of QED we have the quarks, with their various flavours and their three colours; in place of the photon, the quantum of QED, we have the eight colourful gluons.

The two theories are actually very similar. QED of course makes use of the quantum version of James Clerk Maxwell's equations for electrodynamics and the equations for QCD are not so very different from Maxwell's. The important distinction is that whereas the photon, responsible for electromagnetism, is itself electrically neutral, the gluons, which carry the colour force, are themselves colourful and couple to that force; this adds a couple of extra terms to the equations, which are then no longer identical to Maxwell's. The solutions are profoundly affected by the existence of these extra terms.

In the case of QCD we know how to solve the equations at very small distances, that is, when the quarks are separated by a distance much smaller than 10^{-13} cm. (Anything bigger than 10^{-13} cm, about the size of the proton or neutron, is, in particle physics, a huge distance.) Now at very small distances, deep inside the neutron or proton, the effective gluon–quark coupling strength tends to zero in QCD. The quarks, consequently, act as nearly free in the deep interior of the neutron or proton. This notion of 'asymptotic freedom' suggested by the theory has been amply confirmed by experiment over the past twelve years or so.

The coupling strength, which gets very small as the distance between the quarks gets small, likewise grows much larger as the distance increases, until ultimately we cannot follow it by calculation any more; it becomes so large we no longer know how to solve the equations correctly. Nevertheless at large distances we suppose that the coupling strength keeps on increasing so rapidly that, unlike all other interactions, the force does not decline at large distances. If that is true it would result in the confinement of quarks and gluons. Coloured quarks and colourful gluons would be permanently trapped inside 'white' objects like the neutron and proton, and they could never be got out. They could then be detected only indirectly, through experiments on, for example, protons and neutrons, which would act as though made of quarks interacting through gluon forces. In fact that is exactly how neutrons and protons do behave. The indirect detection of quarks, and more recently of gluons, has been carried very far experimentally so that it looks practically impossible now to abandon the notion of a quark. Even the gluon now has very considerable experimental support.

If we are right, indirect evidence is all there ever will be. This of course does not make QCD an unsuitable scientific theory because the predictions are quite definite about how observable objects should behave. Of course some of them still require theoretical work, namely those concerned with the large-distance behaviour.

It should be mentioned that both experimentally and theoretically there is still the possibility that confinement of colour is only approximate, that some small leakage effect permits a detection of individual quarks, and that bulk matter contains a tiny proportion of unconfined quarks. In fact, in one experiment on niobium spheres at Stanford such a result is claimed. If quarks really are sometimes unconfined, that would not only affect fundamental physical theory but lead, most probably, to practical consequences as well. At least one fractionally charged particle would be perfectly stable, because of electric charge conservation, and would have striking chemical properties. It has been debated whether such particles could catalyse thermonuclear fusion. In any case, we can imagine the growth of a prosperous quarkonics industry. However, for the rest of this talk I shall assume that colour confinement is perfect.

We may remark that the old problem of the 'nuclear force' is solved in principle in the sense that we now believe the binding of neutrons and protons to form the nucleus is an indirect consequence of the basic quark–gluon interaction, described in QCD. Such a situation is not at all unprecedented: inter-atomic and inter-molecular forces, so dear to the chemists, are of the same kind; they also are not fundamental, but just indirect consequences of electromagnetism, as treated in QED.

Besides the neutron and proton, and the central nucleus made of them, there are also of course the electrons that surround the nucleus to form the atom. Here we turn to another part of the elementary particle system.

The electron does not have colour and does not feel the nuclear force. But

Table 8.3

Particles and forces

Forces			*QFD* Electro-weak or Flavour force: *Electromagnetic QED*	*QFD* Electro-weak or Flavour force: *Weak*	*QCD Colour force*	*Einstein's theory of Gravitation*
Exchange quanta, the mediators of quantum field theories.		Name	Photon γ	$X^{\pm}$, Z^0	Gluons (8 Colour combinations)	Graviton
		Mass	0	80 90	0	0
		Spin	1	1	1	2
Elementary particles (all spin ½) and forces they experience	charge	Quarks. (3 Colours)				
	$+\frac{2}{3}$, $-\frac{1}{3}$	$\begin{pmatrix} u \\ d \end{pmatrix}, \begin{pmatrix} c \\ s \end{pmatrix}, \begin{pmatrix} t? \\ b \end{pmatrix}, \begin{matrix} \cdots \\ ? \\ \cdots \end{matrix}$	Yes	Yes	Yes	Yes
		Leptons				
	0	ν_e, ν_μ, ν_τ, . . .	No	Yes	No	Yes*
		?				
	−1	e^-, μ^-, τ^-, . . .	Yes	Yes	No	Yes
		Plus antiquarks and antileptons				* Gravitation couples to energy of massless neutrinos.

in a very important sense it has flavour, like the quarks. Just as the u and d quarks are flavour partners, so the electron has a flavour partner, the electron-neutrino, ν_e^0. The electron-neutrino, like the electron, lacks colour and feels neither the nuclear force, nor, since it is electrically neutral, the electromagnetic force; it can pass right through the Earth with very little chance of interacting. That stimulated John Updike to write this poem about it:

COSMIC GALL
by
John Updike

Neutrinos, they are very small.
They have no charge and have no mass
And scarcely* interact at all.
The earth is just a silly ball
To them, through which they simply pass,
Like dustmaids down a drafty hall
Or photons through a sheet of glass.
They snub the most exquisite gas,
Ignore the most substantial wall,
Cold-shoulder steel and sounding brass,
Insult the stallion in his stall,
And, scorning barriers of class,
Infiltrate you and me! Like tall
and painless guillotines, they fall
Down through our heads into the grass.
At night, they enter at Nepal
And pierce the lover and his lass
From underneath the bed—you call
It wonderful; I call it crass.

The neutrinos do of course interact—they can be produced, and so, by a kind of reverse process, they can be detected. They interact through the weak force, which can cause a neutrino to turn into its flavour partner the electron while a neutron changes into a proton. Actually the basic interaction (Fig. 8.3) involves a d quark turning into a u quark; flavour exchange takes place as the electron-neutrino turns into an electron and the d turns into the u. In a quantum field theory context this occurs with the exchange of a particle called X^+; there is, of course also an X^-. These quanta are heavy, electrically charged and predicted to be found in the next few years in experiments just higher in energy than those being done now. They must be found! These are not hidden particles trapped inside matter; they must be observed individually, or else large numbers of us theoreticians will fall on

* The original reads 'And do not interact at all'. This change is made by scientific licence.

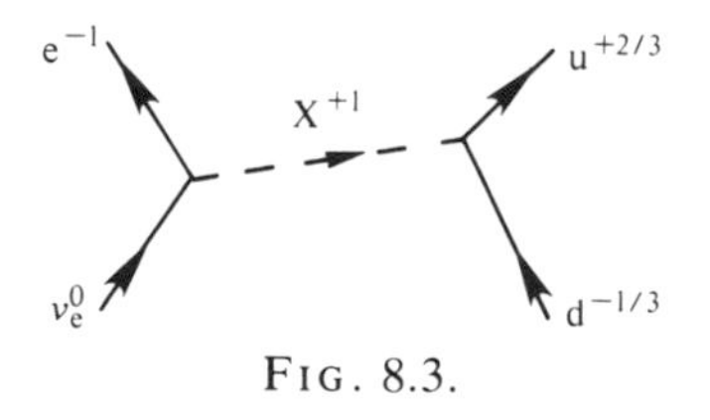

FIG. 8.3.

our fountain pens. (For clarity, I should say there are some misguided colleagues who refer to the X^+ and X^- as W^+ and W^-; this, I'm sure, will stop.)

The electromagnetic and weak forces are 'flavour forces': the electric charge of a particle depends on its flavour; weak forces are flavour exchange forces. Altogether, then, we have a set of flavour forces that we can think of as the electro-weak force. A quantum flavour-dynamics (QFD) has been formulated for the electro-weak interaction; it includes QED and a description of the weak force as well. It has made many successful predictions, for example the existence of a new flavour force, discovered in 1973, in which neutrinos can simply scatter off quarks without exchanging flavour, without turning into electrons. In terms of quantum field theory the neutrino and quark are exchanging another intermediate quantum (of a new weak force) called Z^0 (Fig. 8.4).

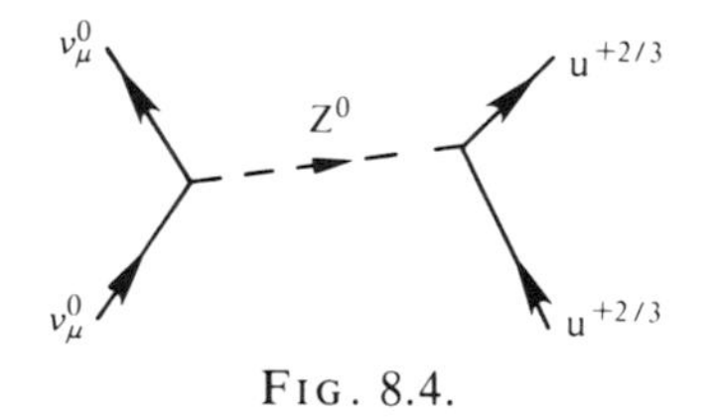

FIG. 8.4.

Again, the Z^0 is supposed to be a real object with a definite mass, about 90 GeV (nearly 100 times the proton mass) and during the coming decade experiments at high energy are supposed to produce it, otherwise we are in deep trouble. In fact the various laboratories are now disputing which one will have the strongest production of the Z^0 quanta in order to be able to use them to study other things; you see how much confidence they have in us theoreticians!

The proposed quantum flavour dynamics has four quanta: the photon, already familiar, and the other three, $X^\pm$ and Z^0, soon to be discovered. The theory, QFD, has certain features connected with the violation of symmetry that may require, if not modification, at least further specification. It cries out, in any case, to be imbedded in a bigger and better scheme. But at low energies it is certainly the right theory, or very close to it, and in 1979 Glashow, Salam, and Weinberg were awarded the Nobel Prize for their contributions to its formulation.

Now let us sum up what we have found so far (Table 8.4). There are the two flavours, electron and electron-neutrino, of what are called, misleadingly, leptons; lepton means light in weight but they are not all light, as we shall see. Then there are two flavours of quark, u and d. (See, quark is a much better name because it is not misleading!) Operating on the colour variable we have colour forces carried by colourful gluons and described by QCD. Working across the table are the flavour forces, carried by the photon, for electromagnetism, and by the $X^{\pm}$ and Z^0 for the weak interaction; all are described within QFD. The quarks and gluons are thought to be confined but all the others must be observable.

Table 8.4
Elementary particles

Leptons	e^-	ν_e^o	
Quarks	$d_R^{1/3}$	$u_R^{+2/3}$	Colour forces (QCD) carried by colourful gluons.
	$d_G^{-1/3}$	$u_G^{+2/3}$	
	$d_B^{-1/3}$	$u_B^{+2/3}$	
	Flavour forces (QFD) carried by photon, $X^{\pm}$, and Z^o.		

Why not stop here? For some reason Nature does not find herself satisfied with this relatively brief list of particles. There are other leptons. The muon, which is just like the electron but 200 times heavier, was discovered at Caltech in 1937, was promptly mistaken by theoreticians for something they had been wanting, and caused tremendous confusion for ten years or so. It is accompanied by its own neutrino, labelled ν_{μ}^{o}.

More recently, in 1975, experiments at SLAC in Stanford, California, discovered the τ-lepton, very heavy, about twenty times the mass of the muon. Its partner the τ-neutrino, ν_{τ}^{o}, has not yet been demonstrated very well experimentally but no doubt exists. There might even be a few more flavours of leptons, although there are cosmological reasons for believing the list may stop here. As I. I. Rabi said about the muon, 'Who ordered that?' We may well ask now 'Who ordered this replication of families? Why does Nature want so many flavours?'

The quarks are in almost exactly the same situation: we have the u and the d with charges of +2/3 and −1/3; we know the strange (s) and charmed (c) quarks also have to be there, the s with charge −1/3 like the d, and the c with charge +2/3, like the u quark. Recently experimentalists discovered, with very little encouragement from theoreticians, another quark, b, with charge −1/3, and many people believe, rather plausibly, that the b will have a partner, t, which will have a charge +2/3. There will then be three pairs of flavours of quarks and three pairs of flavours of leptons.

In addition to all these of course we must remember the antiparticles. In quantum field theory there is always a symmetry between particles and their so-called antiparticles. And of course the particles are the antiparticles of antiparticles; it is just a reciprocal symmetry. Under this symmetry the electric charge turns into its opposite, the mass remains the same, you have to turn left into right and forward-time into backward-time, but I won't dwell on these aspects. For some neutral particles, for example the photon, the antiparticle and particle are one and the same; but in most cases particle and antiparticle are distinct objects. For instance the proton is distinct from the antiproton, the neutron is distinct from the antineutron, the quarks are distinct from the antiquarks. The antiproton is made of three antiquarks, as the proton is made of three quarks, and so on.

If we take ordinary matter and replace electrons by their antiparticles, the positrons, and replace protons and neutrons by their antiparticles we get 'antimatter'. In isolation, antimatter looks very much like matter, but of course if you brought a lump of antimatter and a lump of matter together they would annihilate with a release of energy, most of it finally assuming the form of photons, neutrinos, and antineutrinos.

In principle, complex antimatter objects could be made. The most complex nucleus that I believe has actually been made at a high energy accelerator is the anti-helium nucleus (or anti-alpha particle), which consists of two antiprotons and two antineutrons. With enough time, trouble, and expenditure, you could in principle make an anti-dust mote, or perhaps even an anti-organism. Here is another poem, this time by a physicist, on matter and antimatter.

PERILS OF MODERN LIVING
by
Harold P. Furth

Well up beyond the tropostrata
There is a region stark and stellar
Where, on a streak of anti-matter
Lived Dr. Edward Anti-Teller.

Remote from Fusion's origin,
He lived unguessed and unawares
With all his antikith and kin,
And kept macassars on his chairs.

One morning, idling by the sea,
He spied a tin of monstrous girth
That bore three letters: A.E.C.
Out stepped a visitor from Earth.

Then, shouting gladly o'er the sands,
Met two who in their alien ways
Were like as lentils. Their right hands
Clasped, and the rest was gamma rays.

Now why has Nature not taken the trouble to make a lot of antimatter and put it near us somewhere? If there were a lot of anti-stars in the Galaxy, or a lot of anti-galaxies in a cluster, then when matter and antimatter stars or galaxies collide, as ordinary ones sometimes are observed to do, the annihilation would generate radiation that could be detected. From the lack of such radiation it looks as if, at least in our part of the Universe, there is not much antimatter; perhaps nowhere in the Universe is there much antimatter. An important question is why?

For the first time an answer to that question may now be available. But let me first take up the natural question of why are there so many kinds of elementary particles; this will be the main theme of the rest of this chapter.

One possible and rather obvious answer is that in fact the quarks and leptons are themselves composite. They show absolutely no sign of that so far, but at very high energies they might begin to do so. Until recently, theories that treated them as composite taught us very little, but now somewhat more interesting suggestions are being made, and perhaps we should take the idea more seriously. In the People's Republic of China, under the rule of the notorious Gang of Four (G 4), the belief in compositeness was almost compulsory—not only that but one had to follow an idea attributed to Chairman Mao (based on the thinking of earlier Communist leaders) that underneath each level of reality there exists another level, with the particles being made up of sub-particles in an infinite 'chain of being'. That notion is no longer compulsory in China under the more liberal rule of today's authorities. It might of course be right, but we have no particular reason for thinking so except that it would represent a sort of historical continuity; what has always happened in the past is that apparently fundamental entities turned out to be composed of still smaller objects.

Another possibility is that there is some completely new way of looking at things that will be pointed out by a brilliant young scientist who will show us that we are just asking all the wrong questions. There is a story about Pauli in Heaven, that he was accused of asking the wrong questions. The story was told, while he was still alive, that he died, went to Heaven, and, very impatiently, demanded to see God. After making his way through the bureaucracy he was admitted to the presence of God and immediately asked him why He had made the particles the way He had—in particular why was there a muon? God listened to all of this, naturally with infinite patience, and then delivered a lecture to Pauli on elementary particles. According to the story Pauli is supposed to have nodded, frequently, during the first part of the lecture, even when God said 'The trouble, Pauli, is that you are asking the wrong questions' he still nodded. Eventually, though, as God got into the deeper mathematics of the theory Pauli began to shake his head, indicating that he found something the matter; presently he couldn't contain himself any more and he stood up and yelled 'Aber das ist ganz falsch!' (That's completely false!) And he pointed out the mistake.

Let me now consider at some length a third possibility we can envisage, namely, that we can utilize the particles that appear elementary today, along with others, to construct some sort of unified quantum field theory.

We note first that the theory of QCD and the theory of QFD are very similar mathematically; they belong to the same class of theories. They are called Yang–Mills theories after two people who introduced them to physics in the 1950s; some people call them 'gauge theories', not a very good name because there are actually other gauge theories. QCD is a Yang–Mills theory with perfect colour symmetry, that is, the three colours are treated in an absolutely symmetrical fashion. In QFD this is not the case; Nature is quite unsymmetrical with respect to the flavours. The photon has zero rest mass while the X^+, X^-, and Z^0 are supposed to have gigantic rest masses, of about 80 or 90 GeV; the neutrino is massless, or very nearly so, while the charged leptons have considerable mass. So there is a great deal of asymmetry in the case of flavour and none in the case of colour.

Many people are searching for a unified scheme, a sort of 'Grand Design' that would embrace flavour dynamics and colour dynamics together, as well as new forces. Unification of colour and flavour would take place in an over-all quantum unified dynamics, which would be a Yang–Mills theory with broken symmetry, like the flavour dynamics but much bigger and including colour. The colour portion would be exactly symmetrical but the rest of the symmetry would be largely broken.

This is the point where I must deliver my warning. From now on we are talking about ideas that are not supported by experimental evidence and could be completely wrong. How is such speculation carried out? Perhaps you have already seen this picture (Fig. 8.5), much publicized in 1979, the centenary of Einstein's birth.

The picture shows the master at work on his theory . . . you see how the inappropriate theories are discarded and finally a correct one is found. The most important tool of our trade is the waste basket, into which we throw all the theories which fail to be self-consistent, or fail to be consistent with some well-established body of evidence. The remainder is small, but includes presumably, that which we want to keep.

One problem with unified Yang–Mills theories is that the effective unification between colour and flavour, which have very different coupling strengths, can take place only at a very, very high energy. This is because in quantum field theories coupling strengths change very slowly with energy and in order for weak interactions and strong interactions to come together at some common value of the coupling strengths one has to move over an enormous range of energies, up to something like 10^{15} GeV. The most important question about this kind of work is how can we have the nerve to

FIG. 8.5.

believe that our physical ideas, developed at lower energies where experimental confirmation is possible (up to energies of 10 GeV or so) could possibly be correctly extrapolated over a factor of 10^{14} until we reach 10^{15} GeV. Very unlikely, one would think! Nevertheless people try and the results are very interesting. Another question we might ask is what about experimental tests: between present energies and the unification energies what sort of experimental tests can we have? It is difficult to imagine any way in which even the world budget would allow the construction of machines capable of verifying behaviour at 10^{15} GeV! We are lucky if we can get to 10^3 GeV.

Certain experiments performed without accelerators may help to test consequences of a unified Yang–Mills theory. In particular, two cases have been discussed in previous chapters; let me refer to them again. The first has to do with proton stability. In constructing a unified Yang–Mills theory of colour and flavour, it is possible to stabilize the proton, but it takes a certain amount of effort on the part of the theoretician. It is more natural, in such a theory, for the proton to decay. But of course we all believe, on the basis of extremely sound experimental evidence, even the evidence of our own senses, the evidence of our own existence, that the proton is stable, or nearly so. If the proton decays, but slowly enough, that could be compatible with everything we know.

Now the present upper limit for the decay rate of the proton is near the ridiculously small number of 10^{-30} per year. In a unified Yang–Mills theory of colour and flavour, the proton decays at a rate of about 10^{-32} per year, not very far below the present experimental limit. Therefore it is worthwhile to try to perform experiments to look for proton decay, to improve the present limit, and see if one actually observes proton decay; such an observation would totally change our impression of what the world is like. Such experiments are being organized in the United States (there is one in a silver mine in Utah and one in a salt mine in Ohio) and there may be experiments soon in Europe as well, perhaps in a tunnel under the Alps. These experiments are supposed to search for an occasional decay of a single proton in a thousand or more tons of water or iron.

Another point, very important, follows if there is proton instability. Suppose you also assume that the evolution of the Universe in its very early moments took place under non-equilibrium conditions and that the so-called CP violation, that is the slight asymmetry that has been observed between matter and antimatter, persists up to very high temperature. Then it has been pointed out that one may be able to construct an explanation of why the Universe seems to consist mostly of matter rather than antimatter. The first person to put these three assumptions together and reach this conclusion was Sakharov, the Soviet physicist who is nowadays noted mostly for activities outside physics. He did this work in 1969 when he had already partially turned from physics to other matters; it was ignored for a long time and only recently has the mainstream of theoretical physics taken it up.

It is now a popular idea that maybe the notion of a unified Yang–Mills theory is right, maybe the proton does decay, and maybe that will help to explain why the Universe is asymmetric between matter and antimatter. But, of course, real conviction depends to a great extent on a successful result of the experimental search for proton decay.

The other experiment is also in a mine, this time a gold mine, the Homestake Mine in Lead, South Dakota, where for many years a team has been searching for neutrinos from the Sun, those self-same neutrinos from the Sun about which John Updike wrote his poem, coming from above during the day and from below at night. Not enough solar neutrinos are found to agree with astrophysical and elementary particle theory. Although the errors are still fairly large, the discrepancy may well be significant. How can we explain the result?

One possibility is that the astrophysical theory of the Sun may have some flaws. The most radical suggestion was that made in desperation by my colleague Willy Fowler, that the Sun has gone out! If the thermonuclear reactor in the centre of the Sun had gone out, permanently or temporarily, we would know that immediately by the cessation of the production of neutrinos, but it would take a million years or so for the news to reach the surface and stop the shining of the Sun in electromagnetic radiation. The proposal was that we are somewhere in that million-year period. When the news reaches the surface, the Sun would turn off its electromagnetic radiation and we should be left with a really serious energy crisis! One can of course discuss less drastic modifications of astrophysical theory.

Another possibility is that the experiment is wrong. This is the only team of experimentalists in the world clever enough to detect one atom that has been transformed in a car-load of cleaning fluid, and since nobody else can check their results, they could conceivably be wrong. A clever, but rather ruthless experimental physicist proposed, presumably in jest, to take into the mine a very strong radio-active source and, without notifying the mine authorities, the health authorities, or the miners, place it near the experiment to see if the neutrinos from the source would be properly detected. Needless to say that hasn't been done; instead the same team has constructed an apparatus that is to be exposed to the accelerator at Los Alamos to see if the neutrinos there will set it off in the proper way.

Yet another possibility is that there is something interesting in particle physics that is responsible for the failure to find enough neutrinos. In unified Yang–Mills theories it can very easily happen that the three neutrinos we know about don't have exactly zero rest mass, that is, they don't travel always with the velocity of light. They could have very very small rest-masses, perhaps of about 1 eV, which would have escaped observation. Not only that, they may have so-called 'transition masses', that is to say, probabilities for transforming from one species into another as they move through space. If that is true then the electron-neutrinos emitted by the sun could be transforming themselves into, for example, τ-neutrinos on their way to the Earth. The ν^0_τ would not be detected by the apparatus in the gold mine and in that way one could account for a modest factor of suppression in the number of neutrinos observed, using this very interesting phenomenon of 'neutrino oscillation'. There are now experiments under way at various nuclear reactors to search for neutrino oscillation.

So although the unified theories deal with unification at fantastically high energies, they have a certain number of rather interesting consequences that might be detectable at ordinary energies and might in fact revolutionize our ideas about the Universe.

What about more general unification? Let me say just a few words about that, to summarize it. I have to add one more bit of physics. Each kind of particle has a definite amount of 'spin' angular momentum; in quantum mechanics that is quantized and in the appropriate units the spin is 0, 1/2, 1, 3/2, 2 . . . and so on; an integer or half integer.

The photon, the $X^{\pm}$ and Z^0, and the gluons all have spin 1; the quarks and the leptons all have spin 1/2. In fact all the elementary particles that we have mentioned so far have spin 1/2 or 1.

However, there are reasons to discuss also spin 0 and spin 2. The spin 0 particles come in from spontaneously broken symmetry, discussed by Chris Llewellyn Smith and Abdus Salam in chapters 3 and 5. In theories with spontaneously broken symmetry we try to use exactly symmetrical equations to produce unsymmetrical effects, a very subtle process; the mechanism for achieving this requires the presence of spin 0 particles. In today's theories they are introduced in what I consider to be a rather deplorable way; the spin 0 particles are dragged in, *ad hoc*, with numerous arbitrary parameters being adjusted to explain some of the quantities we observe. I think that can't last. To avoid such arbitrariness in the discussion of these spinless particles some theorists are trying to explain them as bound states of other particles, perhaps new ones, or to find some symmetry principle that requires their existence.

There is also a very important reason for studying spin 2, namely to incorporate Einstein's theory of gravitation, which he liked to call General Relativity, into the framework of quantum mechanics. This requires the graviton, the quantum of gravitation, and because of the nature of Einstein's theory the graviton must have spin 2. It is difficult to find the graviton experimentally because the gravitational coupling is so small for little bits of matter; for a whole planet, of course, there is a sizeable coupling to gravity but an electron, with its tiny mass, couples very weakly to gravity. Consequently the experimental discovery of the graviton must be postponed to a later age. Nevertheless, if you believe both Einstein's theory of gravity and quantum mechanics you must have a graviton; therefore some theorists have engaged in an even more ambitious unification scheme that involves putting gravitation (Einstein's theory) as well as the colour and flavour forces all together in one theory. That theory should include not only all the quanta but also the quarks and leptons as well as the spinless particles we believe we need for spontaneous symmetry breaking. If some or all of these particles are composite, then their fundamental constituents would be described by the theories. The aim is to incorporate all fundamental objects, including the graviton, into one single, truly unified theory. One makes use of 'super-symmetry', a symmetry connecting particles of different spins, to relate the various elementary particles to one another.

The biggest and best such theory to have been found so far is called '$N = 8$ super-gravity' and contains fundamental fields corresponding to:

1 graviton of spin 2;
8 new objects of spin 3/2, called gravitinos;
28 quanta of spin 1;
56 particles and antiparticles of spin 1/2;
and 70 objects of spin 0, possibly useful for symmetry breaking.

The gravitinos of spin 3/2 are a welcome addition from the point of view of simplicity: when we discussed only spins 2, 1, 1/2, and 0 we had an ugly gap at spin 3/2.

Now, if we try to interpret the 28 quanta of spin 1 as including the gluon, the $X^{\pm}$, the Z^0, and the photon, we find that the mathematics is too restrictive. At least the $X^{\pm}$ would have to be left out. Likewise, if we try to interpret all or most of the 56 particles and antiparticles of spin 1/2 as leptons, antileptons, quarks, and antiquarks, we find that we cannot accommodate enough flavours to agree with the observed list.

The $N - 8$ super-gravity theory comes remarkably close to fulfilling our present-day version of Einstein's dream of unifying all the forces of Nature in a single equation. However, it does not quite work if we try to identify today's elemenary particles of spin 1/2 and spin 1 with those of the theory. Either we have to look further for the right unified field theory or else we have to admit that some or all of our 'elementary particles' are not really fundamental.

Let me mention that in super-gravity, or any unified scheme including Einsteinian gravity, the energy of effective unification is even higher than before; we need even more nerve to believe in this kind of theory than we did for the unified Yang–Mills theory unifying just colour and flavour forces; here we are extrapolating from the 10 GeV or so of present experimental knowledge to something like 10^{19} GeV, an even larger factor.

Where do we stand then, in the quest for unity in the description of the elementary particles that make up the world? Some immediate issues, as I see them, are the following.

I would like to know, better than the present theories tell us, what is the right way to look at spontaneous symmetry breaking. If we really have spontaneous symmetry breaking induced by spinless particles, are those introduced into the theory as elementary objects? If so, do they have a reason for being in the theory because of something like supersymmetry or, if not, do they result from the binding of other particles in the theory? I find it very hard to believe that they need to be dragged in *ad hoc* with numerous arbitrary parameters having to be adjusted to fit the data.

What produces this curious replication of quark and lepton flavours? Why do we have not just u and d quarks, but also c and s, b and probably t quarks? Why, for the leptons, do we again have three pairs of flavours: electron and its neutrino, muon and its neutrino, τ and its neutrino? Does the replication, somehow, go on further? We note that cosmologists don't want us to have more than three or four neutrino types.

How do we explain the curious mass spectrum of the quarks and leptons: light ones, medium light ones, heavy ones? Where do these strange mass ratios come from?

We really have, I think, very little idea of the answer to any of these questions. The big questions, of course, relate to unification and to the identification of fundamental entities. Will the attempts at over-all unification or the slightly more modest attempts at unification without gravitation succeed? In a proposed unified theory can we figure out the relation between the elementary objects in the theory and the particles that would appear at present or future experimental energies as elementary? Not only composite objects but other particle-like solutions of the fundamental equation may masquerade as elementary and produce confusion. Nevertheless we still hope to discover a single elegant mathematical equation with a unique structure that will account for the three colours, the right number of flavours, and all the other special features of particle physics. It would be an equation for a giant superfield with many components representing different elementary entities unified by a basic symmetry of Nature. If our efforts succeed, then simplicity will lie not in economy of particles but in economy of principle.

Let me finish by saying that it is a curious thought that we might, conceivably, come to the end of the description. It is much easier to imagine an unending search, level after level of reality, as in Chairman Mao's 'straton' picture. But we can try to envisage the possibility of solving completely the problem of the fundamental physical laws.

What would it be like? We can only describe it operationally, at least those of us who are not philosophers. Operationally it would look like this: with further experimental support for today's theories at present energies, we theoreticians would propose a unified theory of everything, compatible with all the known facts and predicting a number of new ones. Experiments would be done over a reasonable period of time and costing a reasonable sum of money and would confirm the theory. This would continue for a while, with no exceptions being found. (Of course that has never happened before but it is conceivable.) Eventually, there would be a limit to human patience and to the resources that would be expended in trying further to check this successful theory. Humanity would proclaim it to be the final fundamental physical theory!

Glossary and notes

These definitions and notes summarize and, in some cases, expand on explanations given in the main text. The aim is to help the reader to gain a qualitative feeling for unfamiliar terms and concepts. Words which are themselves treated elsewhere in the glossary are written in italics.

Angular momentum measures the tendency of a body to maintain its state of rotational motion. It is a *vector* quantity with a direction along the axis of rotation chosen so that the rotation appears clockwise when viewed in that direction. It is *conserved* and is quantized in units of $\hbar$. (See also *spin*.)

Annihilation. The process in which a particle *antiparticle* disappear and their total *mass* is converted into *energy* or new particles and antiparticles.

Antiparticle. A particle and its antiparticle have certain opposite attributes such as sign of electric *charge*, *magnetic moment*, *flavour* (e.g. strangeness of +1 or −1), *lepton* number, *baryon* number, etc. But *mass*, *spin*, and *lifetime* must be identical. Normal matter is composed of *protons*, *neutrons*, and *electrons*; antimatter would be composed of the corresponding antiparticles: antiprotons, antineutrons, and antielectrons, or *positrons*. The choice of which to call 'particle' is a matter of convenience and the antiparticle of the antiparticle is the particle. In certain cases, for example the *photon* and neutral *pion*, the two conditions are not distinguished; such particles are their own antiparticles.

Baryon. One of the two sub-classes of the *hadrons*. The lightest baryon state is the *proton* which is apparently stable; its *lifetime* is at least 10^{30} years. This stability, which may not be absolute, is associated with a rule called conservation of baryon number, B; baryons and antibaryons have opposite baryon number (+1 or −1) and the total baryon number is a constant, so they can be destroyed (as in *annihilation*) or created only in baryon plus antibaryon pairs. Baryons have an internal structure containing 3 *quarks*, each of baryon number 1/3; antibaryons contain 3 antiquarks of baryon number −1/3.

Big Bang. See *Expansion of the Universe*.

Black hole. A source of gravitational *field* so intense that *photons* (light rays) cannot escape.

Boson. The name given to particles which do not obey the *Pauli exclusion principle*. All bosons have integer *spin* (0, 1, 2, . . .) in units of $\hbar$. Examples are the *photon*, of spin 1, and all *mesons*.

C, Charge-conjugation. The operation which transforms particles into *antiparticles* and vice-versa.

c. The speed of light: 186 000 miles/s or 2.998×10^5 km/s.

Charge. Electric charge is the source of the electromagnetic force. There is only one sort of electric charge and, by convention, that carried by the *proton* is defined as positive while the *electron* carries the anti-charge, equal in magnitude but negative in sign. Electric charge is *conserved* and is quantized, that is it only occurs in Nature in amounts which are integer multiples of the magnitude of the electron's (or proton's) charge, e. The only exceptions occur for the *quarks* which have charges of 2/3 and 1/3 of e, but these, it is believed, are permanently confined within the *hadrons*. The electric charge determines the strength of the electric force between two charge-carrying bodies and it is to the charge that the exchanged virtual *photon* mediating the force *couples*. By analogy the word charge is often used for the equivalent concept in other forces as, for example, in '*colour charge*' for the *strong* force. In particle physics the magnitude of the electron's charge is adopted as the unit of electric charge; in standard units it is 1.602×10^{-19} coulomb.

Chiral. The name given to an approximate hidden symmetry—see chapter 3, page 69.

Colour is an attribute of *quarks* (and has no connection whatever with the normal meaning of the word). There are three varieties of colour (and three anti-colours carried by the antiquarks). It is believed to be the source, or *charge*, of the *strong* force as described by the theory called quantum chromodynamics or QCD.

Conservation law. A quantity is said to be conserved if, within a system free from external interference, it remains constant in time. For example the total electric *charge* (the sum of all positive charges minus the sum of all negative charges) of an isolated system must remain constant.

Cosmic radiation. A continuous rain of particles arriving at the Earth from outer space. A low-energy component originates in the Sun but the more energetic ones, (90 per cent *protons*, 9 per cent helium nuclei, 1 per cent heavier nuclei) have their source in poorly understood astrophysical processes elsewhere in the Galaxy or, for those of highest energy, perhaps in other galaxies.

Coupling. A term used to describe the interaction between the exchange particle of a *field* (e.g. virtual *photon*) and the particle experiencing the force (e.g. *electron*); hence also coupling constant, specifying the strength of the interaction.

CP. A combination of the two transformations C and P carried out one after the other in either order.

Cross-section. A measure of the probability for a given process to take place. The concept is a very simple one: the chance that a random dart-thrower will hit the dart board is proportional to the area of the board, roughly 1600 cm^2, but his chance of hitting the 'bull's-eye' is more than 1000 times less since its area is only a little more than 1 cm^2. Typical cross-sections in particle physics are: 3×10^{-26} cm^2 for the interaction of two high-energy *protons*, which is comparable to their actual geometric 'target area'; or 10^{-37} cm^2 for a 10 GeV *neutrino* to interact with a proton, showing that the neutrino, which can only interact through the weak force, will on averge pass through about 10^{11} protons before anything happens. The standard unit for cross-section is the barn (as in 'barn door') which is 10^{-24} cm^2; a more useful size for particle physics, at least for strong interaction processes, is the millibarn, mb, 10^{-27} cm^2.

Density of states factor. A particle emitted in a reaction may have one of many possible directions in space and a range of possible values of momentum. In *quantum mechanics* the probability for the reaction to take place is proportional to the 'volume' defined in an abstract space, formed from the three dimensions of ordinary space and the three corresponding components of momentum, accessible to each of the particles emitted. This space is called 'phase space', or 'momentum space', and the number of possible states is the accessible volume divided by the minimum 'volume element' in this space which is h^3.

e. The magnitude of the electron's electric charge.

Electron. The electron is a member of the *lepton* class of particles and is the chemically active component of atoms, bound by electrostatic attraction to the central nucleus. It carries a negative electric *charge*, e, of 1.602×10^{-19} coulombs, its *mass* (m_e) is 0.511 MeV/c^2 (or 9.1095×10^{-28} gm), it has a *spin* of $1/2\hbar$, and a *magnetic moment* of -1.00115965241 ($e\hbar/2m_ec$) which agrees, to the last decimal place, with the value expected for a point-like, structureless particle. Because there is no less-massive particle which carries electric charge the *conservation* of electric charge requires the electron to be stable.

Energy. Energy is an agent, or a product, of change; either it is required to make changes occur or it is released when they happen. *Mass* is equivalent to energy, a sort of stored energy with a very high conversion efficiency: $E = mc^2$. All processes of change are in fact transformations between mass and energy brought about through the agency of one of the four basic forces in Nature and the total mass-energy of an isolated system is conserved. The standard scientific unit of measure for energy is the joule; a 100 watt light bulb consumes 100 joules of energy every second. But the joule is too big to be useful in atomic and particle physics where we use instead the *electron-volt, eV*. One eV is the energy acquired by a particle carrying a positive electric *charge*, equal in magnitude to that of an electron, in falling through a potential difference of 1 volt. 1 eV is equivalent to 1.602×10^{-19} joules. In fact, for particle physics, the most convenient units are 10^6 eV (1 MeV) and 10^9 eV (1 GeV).

eV, electron-volt. See *Energy*.

Expansion of the Universe. Measurements of the wavelengths (colour) of light from distant galaxies show the emission spectra characteristic of specific atoms to be shifted towards longer wavelengths; that is, towards the red end of the spectrum. This red shift is believed to be a Doppler effect, like the fall in pitch of a train whistle as the train rushes past and speeds rapidly away. The red shift shows that most of the galaxies are receding from each other, the Universe is expanding. The speed of recession is greatest for the most distant galaxies and is equal to the distance multiplied by the Hubble constant: about 20 km per second per million light-years. Allowing for some uncertainty in the Hubble constant and for the speed of recession to have been somewhat greater in the past, this implies that the matter in the Universe originated from the same place in a 'Big Bang' between 10 and 20 times 10^9 years ago.

Fermi interaction. The first, and very successful, theory of the *weak* force, due to Fermi.

Fermion. The name given to particles which obey the *Pauli exclusion principle*. All fermions have half-integer (1/2, 3/2, 5/2 . . .) *spin* in units of $\hbar$. Examples are the *electron* and *proton* with spin 1/2, and the Δ^{++} baryon state with spin 3/2.

Field. The region of influence of a force; for example the electric field surrounding an electric *charge*. In quantum field theory the force is propagated by exchange of particles: the field quanta (e.g. *photon*, the *quantum* of the electromagnetic field).

Fine structure constant, α. This is the quantity $e^2/\hbar c$, a dimensionless number with the value 1/137.036. Its name derives from the fact that it determines the magnitude of the splitting, or fine structure, of atomic energy levels caused by the *electron's magnetic moment*. In quantum electrodynamics its value is a measure of the strength of the interaction between two electronic *charges* (e) arising from the exchange of a single virtual *photon*. The importance of this number, a dimensionless quantity formed from the three fundamental quantities e, $\hbar$, and c strongly suggests that there must be a fundamental relationship between them.

Flavour. The quality which distinguishes different types of *quark*: up, down, strange, charmed, bottom (or beauty), and (yet to be discovered but strongly anticipated) top (or truth). The term can be extended to include, as additional flavours, the six types of *lepton*: *electron* and electron-*neutrino*, muon and muon-neutrino, tau and tau-neutrino. Flavour and electric charge are related. Quark and lepton flavours and charges can be changed by the *weak* interaction and the electro-weak theory which unites the weak and electromagnetic forces is also called quantum flavour dynamics, or QFD.

Gauge transformation. In an electric field there is no way of defining, no way of measuring, an absolute value for the electrostatic potential; all phenomena are invariant to global changes in the value of the potential. This is a kind of gauge invariance and in the case of electrostatics it can be shown that the law of conservation of electric *charge* is a consequence of this symmetry. In quantum *field* theory there is another form of gauge invariance which has to do with the phase of the functions describing particles and their propagation (the phase determines the state of the wave associated with the propagation of a particle, say 'crest' or 'trough', at a chosen reference point). There is no way to measure an absolute phase and so the theory is required to be invariant to operations which change the phase. In fact the most interesting, and very restricting, form of gauge invariance is not global, but local; that is, invariance is required to phase changes which may differ from point to point and at different times. The establishment of a local gauge symmetry requires the existence of a force, called a 'gauge force' mediated by exchange of 'gauge particles'. The word 'gauge' was originally introduced by Hermann Weyl in the context of a theory requiring invariance to certain changes in the scale of space, like the choice of different 'gauge blocks' used for calibration by machinists. It no longer has this significance in modern theories but the name remains.

Gluon. The *field quantum* or exchange particle mediating the postulated '*colour* force' which binds *quarks* to form *hadrons* and which is the primary form of the *strong* force. Quarks carry one of three colours and antiquarks the corresponding anti-colours. The colour force is propagated by the exchange of coloured gluons: they form a set of eight, corresponding to eight different colour plus anti-colour combinations belonging to an octet multiplet of the symmetry group *SU(3)*. The three colours and three anti-colours can also be combined in a ninth way, to form a singlet of SU(3), but this is 'white', or neutral in the colour charge and so plays no part in the force. The SU(3) symmetry of colour is exact and free gluons would have zero *mass*, however, like quarks, they are believed permanently confined within hadrons. The gluons have spin $1\hbar$, zero electric *charge*, and no electromagnetic or *weak* interaction.

Graviton. The *quantum* of the gravitational *field*. In a quantum field theory of gravity the force would be propagated by the exchange of a graviton, a particle with zero *mass*, zero *charge*, and *spin* $2\hbar$.

h, ħ. Planck's constant $h = 6.626 \times 10^{-34}$ joule seconds or 4.136×10^{-21} MeV seconds. The magnitude of this quantity determines the scale of phenomena for which ordinary mechanics is no longer a good approximation and must be replaced by *quantum mechanics*. It has the same dimensions as *angular momentum* which is quantized in units of $\hbar$, equal to $h/2\pi$.

Hadron. The class of particles which experience the *strong* force. These are the *baryons*, composed of three quarks, and the *mesons*, composed of a quark and antiquark.

Heisenberg's uncertainty principle. See chapter 2, page 39.

Intermediate vector bosons. These are the three *spin* $1\hbar$ particles acting as mediators of the *weak* force: W^0, W^- and $Z°$ (*Note*: in chapter 8, Gell-Mann prefers the letter X to denote the Ws.) In the electro-weak theory of Salam and Weinberg the $W^{\pm}$ and Z^0 masses are predicted to be about 80 GeV/c^2 and 90 GeV/c^2, respectively.

Lepton. The class of point-like particles with *spin* 1/2 $\hbar$ which do not experience the *strong* force. There are three pairs of leptons: *electron* and electron-*neutrino*, muon and muon-neutrino, tau and tau-neutrino and also the corresponding three pairs of antileptons. The reason for three pairs of leptons is not understood, neither is it known why there are three pairs of *quarks* (one of which is still to be discovered) but there are good reasons to expect the same numbers of both lepton and quark pairs. For each of the three families there is a lepton number, L, which appears to be *conserved* just as *baryon* number is conserved.

Lepto-quark. The name sometimes used for the mediators of the postulated electro-nuclear force in the 'Grand Unified Theory' which is an attempt to unify the electromagnetic, *weak*, and *strong* (*colour*) forces. Their mass is supposed to be about 10^{15} GeV, they have spin 1 $\hbar$ and are usually denoted by the letter X (except in chapter 8 where Gell-Mann uses X for the mediators of the electro-weak force).

Lifetime. The spontaneous decay of radioactive nuclei, or of unstable particles, is characterized by the average time before decay or *mean lifetime*. The distribution in actual lifetimes for a large sample of the same state is exponential. If N_0 are present at time zero then there will be a number $N = N_0 e^{-t/\tau}$ after a time t, where τ is the mean lifetime.

Light year. Astronomical unit of distance equal to the distance travelled by a light ray in one year: 9.46×10^{12} km.

Magnetic moment. A rotating body carrying electric *charge*, such as an *electron* orbiting in an atom or a spinning particle or nucleus, generates a magnetic *field* rather like that of an ordinary bar-magnet. The familiar magnetism of iron has its source in the magnetic moment associated with the intrinsic *spin* of the electron. The strength of such a source of magnetism and its direction (pointing in the same or the opposite direction as the spin) are specified by the magnetic moment. The usual unit is the magneton, $e\hbar/2mc$, where m is the *mass* of the particle.

Mass is a measure of inertia, that is, of the reluctance of a body to change its motion; the force required to achieve a given acceleration is proportional to mass. The mass of a particle depends on its speed relative to the observer, thus when a particle's mass is given it is understood that this is the value which would be obtained under conditions (in a 'frame of reference') in which its speed is zero. This is called the 'rest mass', m_0. Then, at a speed v the '*relativistic* mass', m, is given by:

$$m = m_0 / \surd(1 - v^2/c^2).$$

So, as long as v is very much less than c the relativistic mass and the rest mass are the same, which is the case for all situations of interest outside a particle physics laboratory, and the reason that Newtonian mechanics works so well is that c is a very large number. Mass and energy are equivalent; the total energy of a particle is given by $E = mc^2$, where m is the relativistic mass, and E is the sum of the kinetic energy (that associated with the motion) and the rest mass energy $E_0 = m_0c^2$. Thus energy also has inertia; for example, a *photon* of energy ε has a relativistic mass of ε/c^2 and behaves like a particle with this mass in, say, a collision with an electron. The rest mass of the photon is zero ($m_0 = 0$) but there is no physical meaning to a frame of reference in which the photon, or any particle of zero mass, is 'at rest': its speed is always that of light, $v = c$. Mass also features in physics in another, and seemingly quite different way: it is the *charge* for the gravitational force. Thus the inertia of energy also places it under the influence of gravity. The present attempts to build a unified theory of the forces are based on fundamental symmetries which require all particles to be massless (i.e. zero rest mass). In these theories mass is introduced by 'symmetry breaking' mechanisms which are among the least well understood corners of the present scene; the very origin of mass is one of the profoundest mysteries and one which seems unlikely to be solved until the relationship of gravitation to the other forces is properly understood. The standard unit of mass is the kilogram (kg) but in particle physics it is more convenient to use a much smaller unit derived from that of energy: eV/c^2. In these units the mass of the electron, for example, is 0.51 MeV/c^2. In practice the speed of light, c, is often set to unity and then the mass unit is written in eV (or MeV or GeV).

Meson. One of the two classes of *hadrons* (the others are *baryons*) these are particles with integer *spin* (0, 1, 2, . . .) in units of $\hbar$ and are composed of one *quark* and one antiquark. There is no conservation law for mesons; they may be created singly and all are unstable.

Momentum. The product of *mass* and velocity, momentum is a *vector* with the same direction as the velocity. It measures the inertia of a body in motion and the force required to change such motion is proportional to the rate of change of the momentum (equal to the mass times the acceleration). Like *mass-energy* it is a *conserved* quantity with the added requirement that the momentum, say before and after a collision of two billiards balls, must balance in every direction.

Neutrino. A neutral *lepton*. Three varieties are known: the *electron-neutrino*, muon-neutrino, and tau-neutrino. They have *spin* 1/2 $\hbar$ and experience only the *weak* interaction. It is usually assumed that the neutrinos have zero mass but there is no understanding of why this should be and a recent experiment, yet to be confirmed, suggests the mass of the electron-neutrino is between 14 eV/c^2 and 46 eV/c^2. The experimental upper limits on the muon-neutrino and tau-neutrino masses are 0.57 MeV/c^2 and 250 MeV/c^2 respectively.

Neutron. The neutron is the neutral companion of the *proton* with which it forms the nuclei of atoms. It is a *baryon* with *spin* 1/2 $\hbar$ and a *mass* of 939.57 MeV/c^2. It has a size of about 10^{-13} cm, a magnetic moment of -1.913 ($e\hbar/2M_pc$) and its *quark* composition is (udd).

Nucleon. *Neutron* or *proton*.

Numbers. To cope in a concise way with the very large range of magnitudes found in physics numbers are expressed powers of ten, rather than using long strings of zeros. A number written as 10^n is the same as 1 followed by n zeros. Thus $10^2 = 100$; $10^6 =$ 1 000 000; ten thousand million (the approximate age of the universe in years) is 10 000 000 000 or just 10^{10}. The number 299 790 000, the speed of light in metres per second, is 2.9979 multiplied by 10^8, that is 2.9979×10^8. Numbers less than 1 are written using a negative sign in front of the n; so 10^{-2} is 1 divided by 100, that is, 1/100 or 0.01; 10^{-6} is 1/1 000 000 or 0.000 001. The *mass* of the *electron* expressed as a fraction of the *proton* mass is 5.446×10^{-4}.

P. The parity transformation, which is equivalent to reflection in a mirror (plus a rotation).

Pauli's exclusion principle. This principle was first deduced empirically by Pauli from an examination of atomic spectra. Certain transitions are systematically absent, all of them involving energy levels of the atom in which two *electrons* would have been in exactly the same physical state. Pauli concluded that electrons are indistinguishable one from another and, in particular, that two, or more, were forbidden to occupy the same state of motion. The exclusion principle applies to all particles of half-integer *spin*, or *fermions*, but not to particles of integer spin, the *bosons*, any number of which may occupy the same physical state.

Photon. The *quantum* of the electromagnetic *field*. Virtual photons mediate the electromagnetic force; real photons transmit the *energy* of electromagnetic radiation. The photon has zero electric *charge*, spin 1 $\hbar$, zero *mass*, and is always moving with the speed of light, c. If the photon had mass this would result in small departures from the expected fall-off in strength of an electromagnetic field with distance from its source. The current best experimental upper limit to the mass of the photon is 8×10^{-49} gm, or 4×10^{-16} eV/c^2; this astonishingly small number is derived from the analysis of measurements of the shape of the magnetic field of the planet Jupiter, obtained during the Pioneer 10 fly-past in 1973.

Pion. The pion, or π-meson, was the first *meson* state to be discovered and was identified with the exchange particle predicted by Yukawa as mediator of the nuclear force, binding *protons* and *neutrons* in the nucleus. Its exchange, as a virtual state, contributes to this force but the basic *strong* interaction is now thought to be due to the exchange of *coloured gluons* between *quarks*.

Planck mass. The effects of gravity are felt in the macroscopic world of large *masses* and distances. The smallness of Planck's constant ensures that quantum effects are significant only for atomic and sub-atomic phenomena where, on the contrary, gravitational effects are negligible. The gravitational force between two masses at a given separation is proportional to Newton's universal constant of gravitation, G, whose value is 6.67×10^{-8} cm^3/gm s^2. The value of $\hbar$ is 1.05×10^{-27} gm cm^2/s. To obtain a guide to the scale of phenomena at which gravitation and *quantum* effects may 'meet' we can, from G, $\hbar$, and the speed of light c (2.998×10^{10} cm/s), calculate a mass:

$$M_{Pl} = \sqrt{(\hbar c/G)}.$$

This is called the Planck mass and its value is about 2×10^{-5} gm. It is very small and for masses greater than this quantum effects in gravitation are negligible. However, looked at from the viewpoint of particle physics this is the huge mass of 10^{19} GeV/c^2 and warns that when we consider masses, or equivalent energies, approaching this then gravitational effects can no longer be ignored. For time intervals less than $\hbar/M_{Pl}c^2$, that is about 10^{-44} seconds, *Heisenberg's uncertainty principle* allows vacuum fluctuations to reach the Planck mass within regions of space smaller than $c \times 10^{-44}$, or about 10^{-33} cm. Within such minute intervals of space-time both gravitation and quantum effects are expected to be significant.

Planck's constant. See *h*.

Positron. The antiparticle of the *electron*. It has the same *mass* as the *electron*, electric *charge* of equal magnitude but positive in sign instead of negative, spin 1/2 $\hbar$ and *magnetic moment* opposite in sign to the electron's.

Proton. The nucleus of the hydrogen atom and, with the *neutron*, constituent of all atomic nuclei. It is a *baryon* of *mass* 938.28 MeV/c^2 (or 1.67×10^{-24} gm), spin 1/2 $\hbar$, positive electric *charge* equal in magnitude to the *electron's* charge (e), and a *magnetic moment* 2.79 ($e\hbar/2M_pc$). If it were a structureless, point-like particle the magnetic moment would be close to 1. Its size is about 10^{-13} cm and it has three quark constituents: (uud). The experimental upper limit for any difference in magnitude of the proton and electron charges is about $10^{-21} \times e$.

Quantum. A small, discrete quantity, usually of energy or angular momentum. For example, the energy carried by a photon is $\varepsilon = h\nu$, where ν is the frequency of the associated electromagnetic radiation.

Quantum mechanics. The set of rules and procedures for describing and predicting the behaviour of matter. Newtonian mechanics, familiar as a prescription for calculations dealing with everyday phenomena, is strictly an approximation which becomes inadequate at atomic and sub-atomic levels where *Planck's constant*, *h*, is no longer small enough to be ignored. An alternative term sometimes used is 'wave mechanics' because there are mathematical similarities to the equations describing wave propagation.

Quark. Quarks are believed to be the basic constituents of all *hadronic* matter, in particular of the protons and neutrons which constitute the nuclei of atoms. They have *spin* 1/2 $\hbar$, fractional electric *charge*, and five different *flavours* are known. A sixth flavour probably exists since there is good reason to believe they form flavour-pairs like the *leptons*. Quarks also carry one of three *colours*. There is a corresponding set of five, or six, antiquarks. Their attributes are summarized in Table G.1.

Table G1
Quarks and antiquarks

	Quarks				*Antiquarks*			
Flavour pairs	*Colour triplets*			*Electric charge (unit e)*	*Anti-colour triplets*			*Electric charge (unit e)*
up	u_B	u_G	u_R	$+2/3$	$\bar{u}_{\bar{B}}$	$\bar{u}_{\bar{G}}$	$\bar{u}_{\bar{R}}$	$-2/3$
down	d_B	d_G	d_R	$-1/3$	$\bar{d}_{\bar{B}}$	$\bar{d}_{\bar{G}}$	$\bar{d}_{\bar{R}}$	$+1/3$
charmed	c_B	c_G	c_R	$+2/3$	$\bar{c}_{\bar{B}}$	$\bar{c}_{\bar{G}}$	$\bar{c}_{\bar{R}}$	$-2/3$
strange	s_B	s_G	s_R	$-1/3$	$\bar{s}_{\bar{B}}$	$\bar{s}_{\bar{G}}$	$\bar{s}_{\bar{R}}$	$+1/3$
top(?)	t_B	t_G	t_R	$+2/3$	$\bar{t}_{\bar{B}}$	$\bar{t}_{\bar{G}}$	$\bar{t}_{\bar{R}}$	$-2/3$
bottom	b_B	b_G	b_R	$-1/3$	$\bar{b}_{\bar{B}}$	$\bar{b}_{\bar{G}}$	$\bar{b}_{\bar{R}}$	$+1/3$

Relativistic. The term used to refer to a situation in which particle speeds approach the speed of light.

Scattering. In particle physics this describes the deflection in the path of a particle by collision with a nucleus or other particle.

Spin. Most particles spin like a top. This is an intrinsic *angular momentum* as distinct from 'orbital' angular momentum which, for example, may be associated with the motion of an *electron* in an atom or the relative motion of *quarks* inside a *hadron*. Spin can take half-integer (1/2, 3/2, 5/2, . . .) or integer (0, 1, 2, . . .) values of angular momentum (in units of $\hbar$) whereas only integer values are possible for orbital angular momentum.

Strong force. The force responsible for binding *protons* and *neutrons* to form nuclei (sometimes called the nuclear force) and binding *quarks* to form *hadrons* (including the proton and neutron).

SU(3). The name for a particular group of symmetry transformations acting on a set of three entities. Larger ensembles of entities which are also symmetric to the same group of transformations are sometimes referred to as multiplets. Multiplets of the SU(3) symmetry group include ones containing 8 and 10 entities. Examples are the *baryon* multiplets of size 8 and 10 made up from the 3 *quarks* u, d, and s; this symmetry is only approximate. An example of an exact SU(3) symmetry is that of the 3 *colours* of quarks, with the coloured *gluons* of quantum chromodynamics (QCD) forming a multiplet of 8 entities.

T. The operation of reversal in time. This was assumed to be an absolute symmetry of Nature until found to be violated, along with the combined *CP* symmetry, in neutral K-meson (K^0_L) decay.

Temperature. Within an assembly of particles (e.g. atoms or molecules) if there is no net transfer of energy from any one region of the system to any other the assembly is said to be in thermal equilibrium and all regions are at the same temperature. The temperature of an assembly of particles is determined by the internal energy of the system, that is the energy associated with the particle motions and states of vibration or rotation. The connection between temperature and energy is made through Boltzman's constant, k, which has the value of 1.38×10^{-23} joules per kelvin or, in the energy units used in particle physics, 8.62×10^{-5} eV per kelvin.

Torr. A unit of pressure. Normal atmospheric pressure at sea level is 760 Torr.

Units. The basic units of length, mass, and time used in this book are as follows.
length: centimetre, cm; metre, m; kilometre, km; also Ångstrom, Å is 10^{-8} cm, and fermi, fm is 10^{-13} cm.
mass: gram, gm; kilogram, kg (see also *mass*).
time: second, s.

Vector. A quantity with both magnitude and direction. Common examples are wind velocity (55 m.p.h. from the south-west) or force (the tension in a fishing line). In particle physics the word is also used to denote particles of spin 1 with properties under the spatial transformations of rotation and reflection the same as those of a vector in three dimensions.

W. The letter usually used to denote the charged mediators ($W^{\pm}$) of the *weak* force. (Note: in Chapter 8 Gell-Mann uses the letter X instead.) Its mass is predicted to be about 80 GeV/c^2.

Weak force. The force responsible for radioactive decay of nuclei with the emission of an *electron* and an antineutrino (or a *positron* and *neutrino*).

X. See *lepto-quark* (and *W*).

Z. The letter used to denote the neutral mediator of the weak force (Z^0). Its mass is predicted to be about 90 GeV/c^2.

Bibliography

A selection of books and articles for further reading. Most of those in section 5 require familiarity with particle physics or cosmology and contain references to original sources.

1. Books for the general reader.

Calder, N. *Violent Universe*, BBC, London, 1969.
and *The Key to the Universe*, BBC, London, 1977.

Feynman, R. *The Character of Physical Law*, BBC, London, 1965 and MIT Press (paperback) 1967.

Gamow, G. *Mr Tompkins in Paperback*, Cambridge University Press, Cambridge, England, 1969.
(Mysteries of relativity and quantum mechanics are illustrated by Gamow's imaginative construction of a dream world in which the velocity of light is much less, and the Planck's constant much bigger, than in the real world.)

Gamow, G. *The Creation of the Universe*, Viking Press, New York, 1952.
(Although not up to date this is a very readable introduction to the formation of the elements in the early Universe by one of the originators of this idea.)

Goldsmith, M. and Shaw, R. *Europe's Giant Accelerator*, Taylor and Francis, London, 1977.
(Story, with many illustrations, of the construction of the CERN 400 GeV proton accelerator.)

Jungk, R. *The Big Machine*, Charles Scribners, New York, 1968.

Mathews, P. T. *The Nuclear Apple*, Chatto & Windus, London, 1971.

Polkinghorne, J. C. *The Particle Play*, W. H. Freeman, 1979.

Segrè, E. *From X-rays to Quarks—modern physicists and their discoveries*, W. H. Freeman, 1980.

Weinberg, S. *The First Three Minutes*, Andre Deutsch and Fontana, London, 1977.
(An outstanding book on the Big Bang by one of the winners of the 1979 Nobel Physics Prize.)

Weisskopf, V. F. *Knowledge and Wonder—the natural world as man knows it.* (Second edition) MIT Press, Cambridge, Mass. 1979.

Yang, C. N. *Elementary Particles*, Princeton University Press, Princeton, 1961.

2. Introductory books and reviews—a little more technical.

Bath, G. T. (editor) *The State of the Universe*, Wolfson Lectures, 1979, Oxford University Press, Oxford, 1980.

Bondi, H. *Cosmology*, Cambridge University Press, Cambridge, England, 1960.

Einstein, A. *Relativity*, Methuen & Co. Ltd., London, 1979.

French, A. P. (editor), *Einstein, a Centenary Volume*, Harvard University Press, Cambridge, Mass, 1979.

Maglic, B. (editor), *Adventures in Experimental Physics*, World Science Education, Princeton.
Discovery of Parity Violation in Weak Interactions, Vol. 3, p. 93, 1973.
Discovery of Massive, Neutral Vector Mesons, Vol. 5, p. 113, 1976.

Sciama, D. W. *Modern Cosmology*, Cambridge University Press, Cambridge, England, 1971.

3. Recent *Scientific American* articles.

Weinberg, S. *Unified Theories of Elementary Particle Interactions*, July 1974.

Webster, A. *The Cosmic Background Radiation*, August 1974.
Glashow, S. *Quarks with Colour and Flavour*, October 1975.
Nambu, Y. *The Confinement of Quarks*, November 1976.
Wilson, R. R. *The Next Generation of Particle Accelerators*, January 1980.
Jacob, M. and Landshoff, P. *The Inner Structure of the Proton*, March 1980.
Barrow, J. D. and Silk, J. *The Structure of the Early Universe*, April 1980.
Greenberger, D. M. and Overhauser, A. W. *The Role of Gravity in Quantum Theory*, May 1980.
t'Hooft, G. *Gauge Theories of the Forces Between Elementary Particles*, June 1980.
Ekstrom, P. and Wineland, D. *The Isolated Electron*, August 1980.
Geballe, T. H. and Hulm, J. K. *Superconductors in Electric-Power Technology*, November 1980.
Wilczek, F. *The Cosmic Asymmetry Between Matter and Antimatter*, December 1980.

4. Review articles in *Nature*—these usually include references to original publications.

Leader, E. and Williams, P. G. *Unified Gauge Theories of Elementary Particles*, Vol. 257, p. 93, 1975.
Mulvey, J. *The New Frontier of Particle Physics*, Vol. 278. p. 403, 1979.
Marciano, W. and Pagels, H. *Quantum Chromodynamics*, Vol. 279, p. 479, 1979.
Gaillard, M. K. *The Weak Interaction*, Vol. 279, p. 585, 1979.
Georgi, H. *Why Unify?*, Vol. 258, p. 649, 1980.

5. Review articles for physicists, with references to original sources.

(i) *Reviews of Modern Physics*

Weinberg, S. *Recent Progress in Gauge Theories of the Weak, Electromagnetic and Strong Interactions*, **46,** 255, 1974.
Richter, B. and Ting, S. C. C. *Nobel Lectures in Physics 1976*, **49,** 235, 1977.
Jones, L. W. *A Review of Quark Search Experiments*, **49,** 717, 1977.
Pais, A. *Radioactivity's Two Early Puzzles*, **49,** 925, 1977.
Penzias, A. A. and Wilson, R. W. *Nobel Lectures in Physics 1978*, **51,** 417, 1979.
Glashow, S. L., Salam, Abdus and Weinberg, S. *Nobel Lectures in Physics 1979*, **52,** 515, 1980.
Dolgov, A. D. and Zeldovich, Ya. B. *Cosmology and Elementary Particles*, **53,** 1, 1981.

(ii) *Annual Review of Nuclear and Particle Science*

Beg, M. A. and Sirlin, A. *Gauge Theories of Weak Interactions*, **14,** 379, 1974.
Kleinknecht, K. *CP Violation and K^0 Decays*, **26,** 1, 1976.
Schwitters, R. F. and Strauch, K. *The Physics of $e^+ e^-$ Collisions*, **26,** 89, 1976.
Sandford, J. R. *The Fermi National Accelerator Laboratory*, **26,** 151, 1976.
Schramm, D. N. and Wagoner, R. V. *Element Production in the Early Universe*, **27,** 37, 1977.
Greenberg, O. W. *Quarks*, **28,** 327, 1978.
Steigmann, G. *Cosmology Confronts Particle Physics*, **29,** 213, 1979.
(A source for some results quoted by J. Ellis in chapter 6.)
Perl, M. L. *The Tau Lepton*, **30,** 299, 1980.
Goldhaber, G. and Wiss, J. E. *Charmed Mesons Produced in $e^+ e^-$ Annihilation*, **30,** 337, 1980.

Index